DevSecOps in Practice with VMware Tanzu

Build, run, and manage secure multi-cloud apps at scale on Kubernetes with the Tanzu portfolio

Parth Pandit

Robert Hardt

BIRMINGHAM—MUMBAI

3

4

5

Defining and Managing Business APIs   87

Part 2 – Running Cloud-Native Applications on Tanzu

Managing Container Images with Harbor   113

11

Enabling Secure Inter-Service Communication with Tanzu Service Mesh   359

12

Bringing It All Together   383

ways to secure your cloud-native applications and Kubernetes platforms. Lastly, if you are a **DevOps engineer** or a **cloud platform operator**, you will learn how to work with these products for various operations around them.

This book can be read and understood by anyone with a couple of years of experience in the area of building and running modern applications. However, some experience with Kubernetes will help to relate the concepts and challenges better. Additionally, the reader should have some experience working with **command-line interfaces** (**CLIs**) to follow all the hands-on instructions.

What this book covers

Chapter 1, Understanding the Need to Move to a Cloud Platform, sets the context and provides a subject background all about modern applications for the reader.

Chapter 2, Developing Cloud-Native Applications, explains how to develop secure cloud-native applications using predefined application templates containing the required framework and the boilerplate code.

Chapter 3, Building Secure Container Images with Build Service, guides the reader on how to build secure container images in a fully automated manner from the application code or artifacts.

Chapter 4, Provisioning Backing Services for Applications, details how to build a catalog of popular open source tools so that their secure container images can be used as application backend services.

Chapter 5, Defining and Managing Business APIs, is all about how to build and manage secure application APIs.

Chapter 6, Managing Container Images with Harbor, provides instructions on how to store and manage container images with a private container image registry.

Chapter 7, Orchestrating Containers across Clouds with Tanzu Kubernetes Grid, unpacks the details of a multi-cloud Kubernetes platform.

Chapter 8, Enhancing Developer Productivity with Tanzu Application Platform, breaks down how to set up a secure supply chain of applications to build a rapid path to production.

Chapter 9, Managing and Controlling Kubernetes Clusters with Tanzu Mission Control, teaches the reader how to centrally manage a fleet of different flavors of Kubernetes clusters deployed in a multi-cloud environment.

Chapter 10, Realizing Full-Stack Visibility with VMware Aria Operations for Applications, demonstrates how to monitor multi-cloud deployed applications, their Kubernetes platforms, and the underlying infrastructure.

Chapter 11, Enabling Secure Inter-Service Communication with Tanzu Service Mesh, explores how to establish secure communication for microservices deployed on Kubernetes in a multi-cloud/multi-cluster setup.

Chapter 12, Bringing It All Together, summarizes the end-to-end picture of the technology landscape, along with some future predictions for the technologies involved.

To get the most out of this book

The barrier to entry for this book is very low. Anybody with some experience in application development and/or infrastructure operations could use this book to learn about what the Tanzu platform has to offer. However, the reader would get the most out of the book by having some familiarity with Kubernetes and containerized applications. To follow the hands-on instructions given in the book, the reader should also be familiar with CLI commands.

Software/hardware covered in the book	Operating system requirements
Tanzu Application Accelerator	Windows, macOS, or Linux
Tanzu Build Service	
VMware Application Catalog	
API portal for VMware Tanzu	
Spring Cloud Gateway	
Harbor	
Tanzu Kubernetes Grid	
Tanzu Application Platform	
Tanzu Mission Control	
VMware Aria Operations for Applications	
Tanzu Service Mesh	

The step-by-step instructions given in the book are performed on macOS systems. However, most instructions will work on either Linux or Windows machines without modifications.

If you are using the digital version of this book, we advise you to type the code yourself or access the code from the book's GitHub repository (a link is available in the next section). Doing so will help you avoid any potential errors related to the copying and pasting of code.

There are places in the book where we have used some sample applications and Kubernetes configurations. Readers may download them using the GitHub repository provided in the following section.

Download the example code files

You can download the example code files for this book from GitHub at https://github.com/ PacktPublishing/DevSecOps-in-Practice-with-VMware-Tanzu. If there's an update to the code, it will be updated in the GitHub repository.

We also have other code bundles from our rich catalog of books and videos available at https:// github.com/PacktPublishing/. Check them out!

Download the color images

We also provide a PDF file that has color images of the screenshots and diagrams used in this book. You can download it here: https://packt.link/xL5cm.

Conventions used

There are a number of text conventions used throughout this book.

Code in text: Indicates code words in text, database table names, folder names, filenames, file extensions, pathnames, dummy URLs, user input, and Twitter handles. Here is an example: "From there, you can follow the instructions in README.md to run the app locally and deploy it to Kubernetes."

A block of code is set as follows:

```
$ kubectl port-forward -n kubeapps svc/kubeapps 8080:80
Forwarding from 127.0.0.1:8080 -> 8080
Forwarding from [::1]:8080 -> 8080
```

When we wish to draw your attention to a particular part of a code block, the relevant lines or items are set in bold:

```
$ chmod +x gradlew && ./gradlew openApiGenerate   # (this will
create a maven project in-place, use gradlew.bat on Windows)
...
$ mvn spring-boot:run
```

Any command-line input or output is written as follows:

```
kubectl apply -f ./accelerator-k8s-resource.yaml
```

Bold: Indicates a new term, an important word, or words that you see onscreen. For instance, words in menus or dialog boxes appear in **bold**. Here is an example: "From the TAP GUI's **Create** page, let's click the **Choose** button underneath the **Hello Fun** Accelerator."

> **Tips or important notes**
> Appear like this.

Get in touch

Feedback from our readers is always welcome.

General feedback: If you have questions about any aspect of this book, email us at customercare@packtpub.com and mention the book title in the subject of your message.

Errata: Although we have taken every care to ensure the accuracy of our content, mistakes do happen. If you have found a mistake in this book, we would be grateful if you would report this to us. Please visit www.packtpub.com/support/errata and fill in the form.

Piracy: If you come across any illegal copies of our works in any form on the internet, we would be grateful if you would provide us with the location address or website name. Please contact us at copyright@packt.com with a link to the material.

If you are interested in becoming an author: If there is a topic that you have expertise in and you are interested in either writing or contributing to a book, please visit authors.packtpub.com.

Share Your Thoughts

Once you've read *DevSecOps in Practice with VMware Tanzu*, we'd love to hear your thoughts! Scan the QR code below to go straight to the Amazon review page for this book and share your feedback.

https://packt.link/r/1-803-24134-9

Your review is important to us and the tech community and will help us make sure we're delivering excellent quality content.

Download a free PDF copy of this book

Thanks for purchasing this book!

Do you like to read on the go but are unable to carry your print books everywhere?

Is your eBook purchase not compatible with the device of your choice?

Don't worry, now with every Packt book you get a DRM-free PDF version of that book at no cost.

Read anywhere, any place, on any device. Search, copy, and paste code from your favorite technical books directly into your application.

The perks don't stop there, you can get exclusive access to discounts, newsletters, and great free content in your inbox daily

Follow these simple steps to get the benefits:

1. Scan the QR code or visit the link below

https://packt.link/free-ebook/9781803241340

2. Submit your proof of purchase
3. That's it! We'll send your free PDF and other benefits to your email directly

Part 1 –
Building Cloud-Native
Applications on the
Tanzu Platform

This part includes the details of different Tanzu tools that can be useful to enhance developer productivity and the security of the software supply chain for cloud-native application development.

This part of the book comprises the following chapters:

1

Understanding the Need to Move to a Cloud Platform

Welcome! If you have any connection to delivering software at scale in the enterprise, we hope this book will be beneficial to you. In this first chapter, we'll establish the context in which a product such as VMware Tanzu could emerge. We'll do this with a walk-through of the evolution of enterprise software over the last 20 years covering the major milestones, inflection points, and problems that surfaced along the way.

Then, once we have established the necessary context, we can give a 10,000-foot overview of Tanzu covering the tools, benefits, and features. And finally, we'll set you up for success by covering some of the technical prerequisites you should have in place if you want to engage with any of the hands-on content in this book.

In this chapter, we will cover the following:

- The challenges of running a software supply chain
- The emergence of the cloud and containers
- Kubernetes
- Outcome-driven approach
- The need for VMware Tanzu

The challenges of running a software supply chain

VMware Tanzu is a modular software application platform that runs natively on multiple clouds and is geared toward important business outcomes such as developer productivity, operator efficiency, and security by default. If you are looking for a hands-on detailed treatment of VMware Tanzu, you won't be disappointed.

However, before diving into the platform's components, it may help to understand some history and background. If you're reading this, there's a good chance you participate in the coding, designing, architecting, operating, monitoring, or managing of software. However, you may not have considered that you are participating in a *supply chain*.

According to Adam Hayes in his Investopedia article, *The Supply Chain: From Raw Materials to Order Fulfillment*, a supply chain "*refers to the network of organizations, people, activities, information and resources involved in delivering a product or service to a consumer.*"

When a piece of software makes the journey from a developer's workstation to an end user, that's as much of a supply chain as when Red Bull and ramen noodles make the trek from raw ingredients to a production facility to a warehouse to the neighborhood grocery store.

Every supply chain has its own set of challenges, and software supply chains are no exception. Most software written today consists of libraries and frameworks containing millions of lines of open source software developed by people who are essentially anonymous and whose motivations are not entirely clear.

Much of that software changes hands many times as it moves from an open source repository to the developer, to source control, to building and packaging, to testing, to staging, and finally, to running in production. Furthermore, the infrastructure on which that software runs is often open source as well, with a worldwide community of hackers working to identify vulnerabilities in the operating systems, network protocol implementations, and utilities that make up the dial tone that your software runs on. This ecosystem presents an enormous surface area for bad things to happen.

For further reading on real-world examples of what can go wrong with software supply chains, I'd recommend a quick search of the web for the 2020 SolarWinds incident or the 2021 emergence of Log4Shell (CVE-2021-44228). The authors of this book, in their capacity as Tanzu solution engineers, have seen first-hand the impact software supply chain issues can have across the financial, government, telecom, retail, and entertainment sectors.

The emergence of the cloud and containers

Happily, the tech industry has begun to coalesce around some key technologies that have come a long way in addressing some of these concerns. The first is what is known colloquially as *The Cloud*. Enterprises with a big technology footprint began to realize that managing data centers and infrastructure was not the main focus of their business and could be done more efficiently by third parties for which those tasks were a core competency.

Rather than staff their own data centers and manage their own hardware, they could rent expertly managed and maintained technology infrastructure, and they could scale their capacity up and back down based on real-time business needs. This was a game-changer on many levels. One positive outcome of this shift was a universal raising of the bar with regard to vulnerable and out-of-date software on the internet. As vulnerabilities emerged, the cloud providers could make the *right* thing to do the *easiest* thing to do. Managed databases that handle their own operating system updates and object storage that is publicly inaccessible by default are two examples that come immediately to mind. Another outcome was a dramatic increase in deployment velocity as infrastructure management was taken off developers' plates.

As *The Cloud* became ubiquitous in the world of enterprise software and infrastructure capacity became a commodity, this allowed some issues that were previously obscured by infrastructure concerns to take center stage. Developers could have something running perfectly in their development environment only to have it fall over in production. This problem became so common that it earned its own subgenre of programmer humor called *It Works on My Machine*.

Another of these issues was unused capacity. It had become so easy to stand up new server infrastructure, that app teams were standing up large (and expensive) fleets only to have them running nearly idle most of the time.

Containers

That brings us to the subject of *containers*. Many application teams would argue that they needed their own fleet of servers because they had a unique set of dependencies that needed to be installed on those servers, which conflicted with the libraries and utilities required by other apps that may want to share those servers. It was the happy confluence of two technical streams that solved this problem, allowing applications with vastly different sets of dependencies to run side by side on the same server without them even being aware of one another.

Container runtimes

The first stream was the concept of cgroups and kernel namespaces. These were abstractions that were built into the Linux kernel that gave a process some guarantee as to how much memory and processor capacity it would have available to it, as well as the illusion of its own process space, its own networking stack, and its own root filesystem, among other things.

Container packaging and distribution

The second was an API by which you could package up an entire Linux root filesystem, complete with its own unique dependencies, store it efficiently, unpack it on an arbitrary server, and run it in isolation from other processes that were running with their own root filesystems.

When combined, developers found that they could stand up a fleet of servers and run many heterogeneous applications safely on that single fleet, thereby using their cloud infrastructure much more efficiently.

Then, just as the move to the cloud exposed a new set of problems that would lead to the evolution of a container ecosystem, those containers created a new set of problems, which we'll cover in the next section about Kubernetes.

Kubernetes

When containers caught on, they took off in a big way, but they were not the be-all-and-end-all solution developers had hoped for. A container runtime on a server often required big trade-offs between flexibility and security. Because the container runtime needed to work closely with the Linux kernel, users often required elevated permissions just to run their containers. Furthermore, there were multiple ways to run containers on a server, some of which were tightly coupled to specific cloud providers. Finally, while container runtimes let developers start up their applications, they varied widely in their support for things like persistent storage and networking, which often required manual configuration and customization.

These were the problems that Joe Beda, Craig McLuckie, and Brendan Burns at Google were trying to solve when they built *Kubernetes*. Rather than just a means of running containerized applications on a server, Kubernetes evolved into what Google Distinguished Developer Advocate Kelsey Hightower called *"a platform for building platforms."* Kubernetes offered many benefits over running containers directly on a server:

- It provided a single flexible declarative API for describing the desired state of a running application – *9 instances, each using 1 gigabyte of RAM and 500 millicores of CPU spread evenly over 3 availability zones*, for example

- It handled running the instances across an elastic fleet of servers complete with all the necessary networking and resource management

- It provided a declarative way to expose cloud-provider-specific implementations of networking and persistent storage to container workloads

- It provided a framework for custom APIs such that any arbitrary object could be managed by Kubernetes

- It shipped with developer-oriented abstractions such as Deployments, Stateful Sets, Config Maps, and Secrets, which handled many common use cases

Many of us thought that perhaps Kubernetes was the technological advance that would finally solve all of our problems, but just as with each previous technology iteration, the solution to a particular set of problems simply exposes a new generation of problems.

As companies with large teams of developers began to onboard onto Kubernetes, these problems became increasingly pronounced. Here are some examples:

- Technology sprawl took hold, with each team solving the same problem differently

- Teams had their own ops tooling and processes making it difficult to scale operations across applications

- Enforcing best practices involved synchronous human-bound processes that slowed developer velocity

- Each cloud provider's flavor of Kubernetes was slightly different, making multi-cloud and hybrid-cloud deployments difficult

- Many of the core components of a Kubernetes Deployment – container images, for example – simply took existing problems and allowed developers to deploy vulnerable software much more quickly and widely than before, actually making the problem worse

- Entire teams had to be spun up just to manage developer tooling and try to enforce some homogeneity across a wide portfolio of applications

- Running multiple different applications on a single Kubernetes cluster requires significant operator effort and investment

Alas, Kubernetes was not the panacea we had hoped it would be; rather, it was just another iteration of technology that moves the industry forward by solving one set of problems but inevitably surfacing a new set of problems. This is where the Tanzu team at VMware comes into the picture.

Outcome-driven approach

The Tanzu team at VMware came into existence just as Kubernetes was hitting its stride in the enterprise. VMware was poised for leadership in the space with the acquisition of Heptio, which brought deep Kubernetes knowledge and two of the original creators. It also acquired a well-honed philosophy of software delivery through the acquisition of Pivotal. The Tanzu team continues to deliver a thoughtful and nuanced Kubernetes-based application platform focused on meaningful business outcomes that were important to customers.

There is no doubt that many mature Tanzu customers were facing some of the problems with Kubernetes mentioned in the last section, but they were also focused on some key outcomes, such as the following:

- Maximizing developer velocity, productivity, and impact

- Maximizing operator efficiency

- Operating seamlessly across the data center and multiple cloud providers

- Making software secure by default

These were the outcomes Tanzu set out to achieve for customers, and in the process, they would take on many of the issues people were running into with Kubernetes.

The Tanzu portfolio would take an outcome-driven approach to deliver an opinionated Kubernetes-based cloud-native application platform that was optimized for operator efficiency, developer productivity, seamless multi-cloud operation, and application security. That is the platform that we'll cover in this book.

The need for VMware Tanzu

Companies with a large development footprint that were looking at or actively using Kubernetes faced two sources of resistance. First, they faced a myriad of problems running Kubernetes at scale across multiple teams. These are the problems listed at the end of the Kubernetes section of this chapter. Second, these teams were under pressure to use technology to deliver meaningful outcomes. This meant developers needed to be operating at their full potential with minimum friction, and operators needed to be able to scale across multiple teams with a unified set of tools. This is the outcome-driven approach described in the previous section. VMware Tanzu is a portfolio of tools geared specifically toward addressing both sets of needs for software teams.

This diagram highlights where VMware Tanzu fits both as the next iteration of software platforms on Kubernetes, as well as the integrated toolkit enabling world-class outcomes from software development, operations, and security engineers:

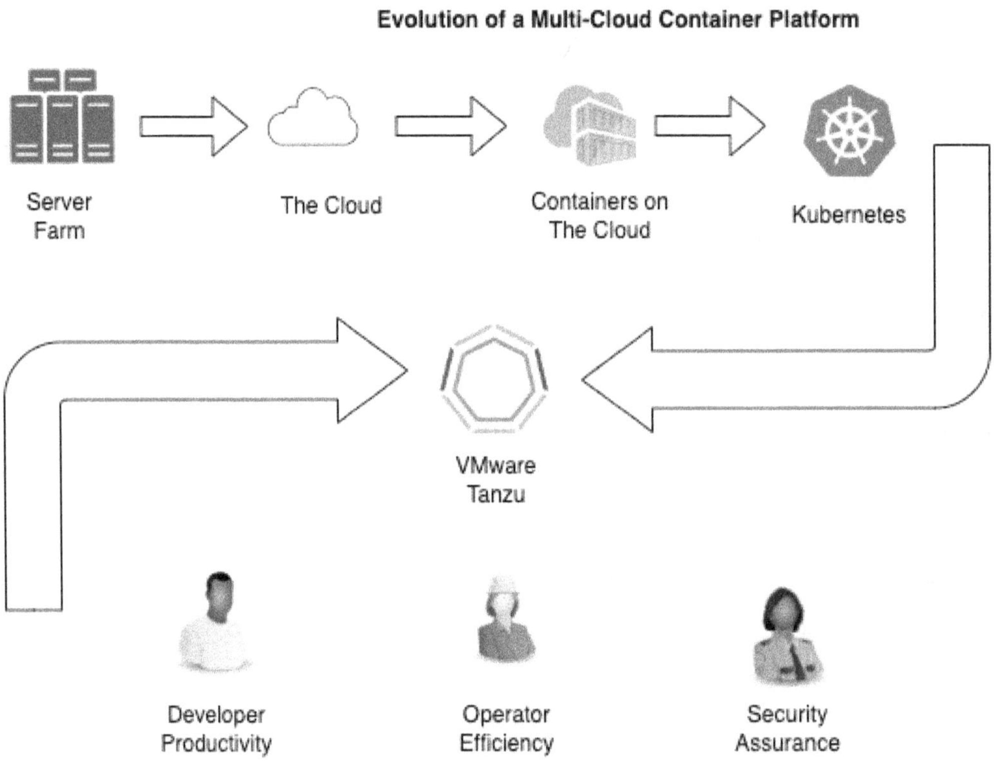

Figure 1.1 – VMware Tanzu in context

Now that we've established the context that prompted the creation of VMware Tanzu, we're ready to describe the product itself.

Features, tools, benefits, and applications of VMware Tanzu

It is a combination of open source, proprietary, and **Software as a Service (SaaS)** offerings that work together to enable the outcomes that are important to software teams. These tools are tightly integrated and give developers a single toolset geared toward delivering important outcomes and running Kubernetes at scale. These tools and applications fall into three broad groups.

Build and develop

Tools that enable developers to efficiently and reliably develop and build software go in this group. This includes *Application Accelerator for VMware Tanzu, VMware Tanzu Build Service, VMware Application Catalog*, and *API portal for VMware Tanzu*.

Run

This group contains the tools and applications to efficiently deliver and run applications on an ongoing basis. It includes *Harbor, Tanzu Kubernetes Grid*, and *Tanzu Application Platform*.

Manage

The final group contains tools for the management of applications and the platform itself. It includes *Tanzu Mission Control, VMware Aria operations for Applications*, and *Tanzu Service Mesh*.

Prerequisites

Now that we've laid out the *why* and the *what* of VMware Tanzu, we're ready to get our hands dirty solving some real-world problems. This book is geared toward software professionals, and there are some tools and concepts that this book assumes you know about. Don't worry if you're not strong across all of these areas as each chapter will walk you through step by step.

The Linux console and tools

You can follow along with most chapters in this book using a Windows machine, but experience dictates that things work much more smoothly if you use a Mac or a Linux workstation. There are numerous options available for Windows users, including virtual machines, dual-booting into Linux, or working from a cloud-based virtual machine. This book assumes you are comfortable with navigating a filesystem, finding and viewing files, and editing them with an editor such as Vim or nano.

Docker

This book is heavily geared toward containers. The primary way to interact with APIs that build and run containers is with the Docker CLI. You will need both a Docker daemon and the Docker CLI to work through some of the chapters in this book. It assumes that you are comfortable listing container images as well as running containers.

Kubernetes

Kubernetes is at the core of the Tanzu portfolio. This book assumes you can stand up a Kubernetes cluster locally or on a public cloud provider. It also assumes that you are comfortable with the kubectl CLI to interact with a Kubernetes cluster. Finally, you should be able to read YAML Kubernetes manifests.

Workstation requirements and public cloud resources

Some of the tools discussed in this book can be run locally on your workstation, while others are better suited to the public cloud. Others require only a web browser and a pre-existing Kubernetes cluster.

If you want to run Tanzu Kubernetes Grid locally, a minimum of 32 gigabytes of RAM is strongly recommended. You may find that other tools, such as Tanzu Application Platform or Harbor, run best on a Kubernetes cluster provided by a public cloud provider. I highly recommend using the providers' built-in budgeting tools to make sure that you don't rack up an unexpected bill.

Now that you know what you'll need, I encourage you to find a topic of interest and dive in. All Packt books are organized with self-contained chapters so you can work through the book front-to-back or jump straight to the topics that interest you.

Summary

In this chapter, we gave a brief history of enterprise software and how its evolution set the stage for VMware Tanzu. Then, we gave a high-level overview of the Tanzu product itself. And finally, we covered some technical requirements to set you up for success in the hands-on exercises.

With all that in place, you're free to jump to any chapter that piques your interest; they're all self-contained and can be read on their own. Or, if you're going in order, we'll start at the beginning of the software development process. Application Accelerator for VMware Tanzu is a tool for bootstrapping a new application with all of the enterprise standards and approved libraries built in, giving you a more uniform portfolio of apps and preventing software teams from repeating past mistakes. Regardless of your approach to the book, we hope you enjoy the material, and we wish you great success putting it into practice.

2
Developing Cloud-Native Applications

A common problem across large enterprises is technology sprawl. When starting a new development endeavor, considerable day-0 cycles are wasted choosing the right technology and bad decisions can lead to ongoing operational headaches for many years.

This chapter will first focus on **Application Accelerator for VMware Tanzu**, or *App Accelerator* for short, a tool that addresses the day-0 problem of standing up a new project. It allows developers to pull down preconfigured application templates that already pass muster with enterprise architecture, have all the corporate security opinions and safeguards built in, include all the corporate standard libraries, use a corporate-standard format and layout, and allow for the configuration of common properties (Git repos, databases, API schemas, etc.)

Then, we'll touch on a couple of development frameworks that currently make up a significant portion of software running in the enterprise: **Spring Boot** and **.NET Core with Steeltoe**.

In this chapter, we're going to cover the following topics:

- The business needs addressed by *App Accelerator*
- Technical requirements
- Overview of *App Accelerator*
- Getting started with *App Accelerator*
- Advanced topics on *App Accelerator*
- Day-2 operations with *App Accelerator*
- Cloud-native development frameworks under the *Tanzu* umbrella

The business needs addressed by App Accelerator

Before we jump into the actual product installation, let's take a moment to think about the life of an enterprise application developer. Let's assume *Cody* is a developer in our organization.

Today, Cody has been tasked with kicking off a new service that is expected to generate significant revenue for the company. If we were to ask him to list a few of the things that could hamper his ability to deliver that service, we might get a list such as this:

- Finding the right application framework for the task at hand
- Importing the necessary libraries into that framework
- Importing all the corporate-mandated libraries
- Resolving any version incompatibilities introduced in the previous two points
- Attesting that all of the corporate security standards are being adhered to in this project
- Making all the imported libraries work together
- Researching and importing all the corporate coding conventions, project structure, documentation layout, and so on
- Waiting for review to ensure that all the standards have been interpreted and implemented correctly

As a software developer, maybe you relate to some of Cody's concerns:

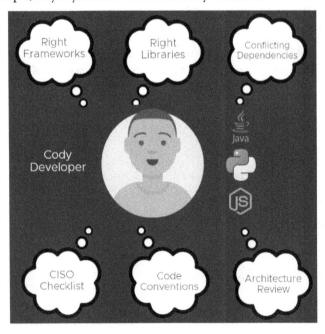

Figure 2.1 – Cody the developer

This is an enormous burden on Cody and even more so for a new developer onboarding onto a project. Much of this work must be repeated every time a new project gets created, it will inevitably result in configuration drift as different developers interpret standards differently, and it constitutes a considerable expenditure of time and effort before a single line of value-delivering code gets written.

Now, let's shift our perspective to that of an enterprise architect. Let's call her *Alana*. She's tasked with optimizing the time to value for developers, setting the technical direction, and creating and enforcing a set of standards that balances consistency and security with developer productivity and satisfaction. She has a very similar list of hindrances that prevent her from maximizing her developers' performance:

- New developers must learn about the quirks of each project before they can become productive
- The process of manually reviewing each component to ensure that all the corporate standards are being adhered to becomes a bottleneck
- Developers gloss over or ignore standards that they don't understand or choose not to implement
- Developers use outdated and unpatched libraries
- Developers use unvetted and unapproved technologies and frameworks
- Job satisfaction for developers suffers due to repetitive, tedious, and error-prone steps in the **Software Development Lifecycle (SDLC)**
- There are few, if any, assumptions she can make about the entire portfolio such that it can be centrally managed
- Attempts at cross-project concerns such as automated scanning, deployment, or backups are hindered by the heterogeneity of the portfolio

If you're focused more on the *Ops* in *DevOps*, perhaps some of Alana's concerns in this visualization resonate with you:

Figure 2.2 – Alana, our enterprise architect

Now that we know why we're here, let's dig into the details of App Accelerator and start addressing Cody's and Alana's pain points.

Technical requirements

App Accelerator is a subcomponent of **Tanzu Application Platform** (**TAP**), which is covered in *Chapter 8, Enhancing Developer Productivity with Tanzu Application Platform*. To get started with App Accelerator, we'll first need to install a Kubernetes cluster and then layer TAP over that. You're free to use whichever Kubernetes distribution you prefer. If you'd like some guidance around standing up a cluster, the appendix at the end of the book gives several options for getting Kubernetes up and running.

Once you have a running Kubernetes cluster, you'll need to jump briefly to *Chapter 8, Enhancing Developer Productivity with Tanzu Application Platform*, where we walk through the installation of TAP. That chapter walks through a more complex end-to-end use case, while this chapter focuses solely on the Application Accelerator component.

Depending on your Kubernetes cluster, your TAP GUI may be at a local address such as `http://localhost:8081`, or possibly at an ingress domain you set up such as `http://tap-gui.example.com`.

When you open a browser to the TAP GUI, you should see something similar to this screenshot:

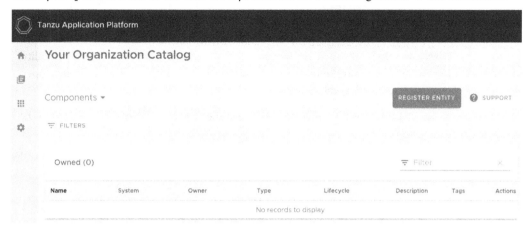

Figure 2.3 – Screenshot of the Tanzu Application Platform GUI

When you click on the **Create** link in the left-hand menu bar, you'll see the default application accelerators that ship with the TAP GUI:

Figure 2.4 – Screenshot of the empty Create section

Congratulations! If you've made it this far, you are ready to start using App Accelerator!

Overview of App Accelerator

App Accelerator is a tool that Alana uses to provide Cody and his peers with a vetted, approved, preconfigured jumping-off point in their language or framework of choice. With App Accelerator, Alana can handle repetitive, low-value tasks such as choosing a technology and ensuring its interoperability and compatibility with other apps and enterprise standards. This frees up Cody and his developer peers to deliver business value with great software.

App Accelerator consists of several APIs and interfaces geared specifically to either Alana the architect's or Cody the coder's persona:

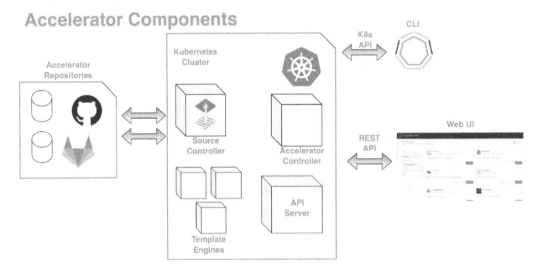

Figure 2.5 – App Accelerator architecture

Alana starts with a Git repo containing the skeleton of a project. In addition to boilerplate code, pre-approved libraries, and standardized documentation and testing structure, this project contains a special YAML file called, by convention, `accelerator.yaml`. This file contains the details of how consumers of the template application (called an *accelerator*) can customize the app template. Running inside Alana and Cody's Kubernetes cluster is a controller for a **Custom Resource Definition (CRD)**, `accelerator.apps.tanzu.vmware.com/v1alpha1`. Alana deploys an instance of the `accelerator` custom resource, and the controller becomes aware of Alana's Git repo, making it available to Cody:

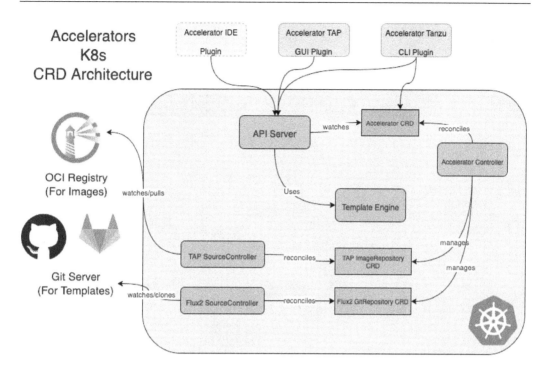

Figure 2.6 – Accelerator CRDs

App Accelerator presents a different view to Cody. He opens up a UI containing details of all of the accelerators available to him. Alana has provided a wide array of languages and frameworks to give developers the flexibility to use the right tool for the job without having to reinvent the wheel with regard to choosing libraries, ensuring compatibility, and navigating the maze of corporate governance.

Cody can search by tags and preview the contents of the various accelerators. Many accelerators will prompt Cody for certain inputs and customize the project accordingly. This is done via the **Template Engine** component in the preceding architectural diagram. App Accelerator provides a rich set of transformations so that a small piece of input from Cody, *what sort of database do you want to use?*, for example, can result in complex configuration options and libraries being preconfigured. The idea is to abstract as much complexity as possible away from the developer to maximize reuse and developer productivity.

Now that we understand how App Accelerator works, let's get started with our own installation.

Getting started with App Accelerator

If you worked through the Technical Requirements section at the beginning of this chapter, you should have a running instance of Tanzu Application Platform and you're ready to begin exploring App Accelerator.

Exploring App Accelerator

When we installed App Accelerator, we got a set of out-of-the-box accelerators that enable developers to hit the ground running with enterprise-ready application templates, like the ones shown as follows:

Figure 2.7 – Application accelerators installed

You can click on the **CHOOSE** button for any of these accelerators and you'll navigate to a page where you can supply any of the required parameters, as well as explore the project template and preview its files. Once you've chosen the accelerator that meets your needs, you can fill in the required parameters and click on **Next Step** and **Create**. At that point, you'll be able to download a ZIP file containing the beginnings of your next great app with all your company's conventions and best practices baked in!

Downloading, configuring, and running App Accelerator

From the TAP GUI's **Create** page, let's click on the **CHOOSE** button underneath the **Hello Fun** accelerator. Let's change the name to hello-fun-tanzu-devsecops and make sure **Kubernetes deployment and service** is selected under **deploymentType**. Then, we click **Next Step | Create | Download Zip File**. This will download a ZIP file that you can unzip, run locally, and deploy to your Kubernetes cluster.

Use your favorite zip tool to unzip the archive and then open the project in your favorite IDE. From there, you can follow the instructions in README.md to run the app locally and deploy it to Kubernetes.

Once you are comfortable downloading, configuring, and running the accelerator, you're ready to move to some more advanced tasks and topics.

Advanced topics on App Accelerator

Let's say you're one of the aforementioned enterprise architects or security engineers who wants to delight your developer colleagues with a panoply of choices from the latest and greatest technologies while baking in all your hard-learned lessons around tech longevity, best practices, and security. How would you go about creating your own app accelerator? Let's do that next. Let's implement a custom app accelerator.

Let's start by logging into GitHub and forking this book's GitHub project. Visit `https://github.com/PacktPublishing/DevSecOps-in-Practice-with-VMware-Tanzu` and click **Fork**. Now, you have your own copy of the code repo in your GitHub account. Your copy of the App Accelerator is located at `https://github.com/<your-username>/DevSecOps-in-Practice-with-VMware-Tanzu/tree/main/chapter-02/openapi-accelerator`.

This accelerator takes as input an OpenAPI 3.0 specification. It then outputs a project capable of building an entire API skeleton around that specification and a SwaggerUI frontend to test it out with. How's that for best practices?

Once you've forked the accelerator project, let's tell the controller running in Kubernetes that we have another app accelerator that we'd like to make available. Let's clone the accelerator locally:

```
git clone https://github.com/<your-username>/DevSecOps-in-
Practice-with-VMware-Tanzu.git
```

Next, navigate to the project and open the `chapter-02/openapi-accelerator/accelerator-k8s-resource.yaml` file in a text editor. Modify the `spec.git.url` (highlighted) property to point to your fork of the project, as shown in the following screenshot:

Figure 2.8 – accelerator-k8s-resource.yaml (point it to your fork of the Git repo)

Then, you can tell the App Accelerator controller about your new accelerator by applying `accelerator-k8s-resource.yaml`:

```
kubectl apply -f ./accelerator-k8s-resource.yaml
```

Now, you should be able to return to the TAP GUI's **Create** page, refresh it, and see your OpenAPI 3.0 Spring Boot accelerator ready to go:

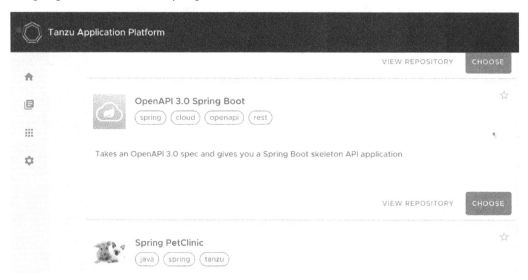

Figure 2.9 – Custom OpenAPI 3.0 Spring Boot accelerator in the TAP GUI

From here, you could use the accelerator to generate a Spring Boot project customized with your own OpenApi 3.0 specification:

1. Click on the accelerator's **CHOOSE** button.

2. Give it a clever name.

3. Use the default (provided) *OpenApi spec* or paste in one of your own. There's a good one here: `https://raw.githubusercontent.com/PacktPublishing/DevSecOps-in-Practice-with-VMware-Tanzu/main/chapter-02/petstore.yaml`:

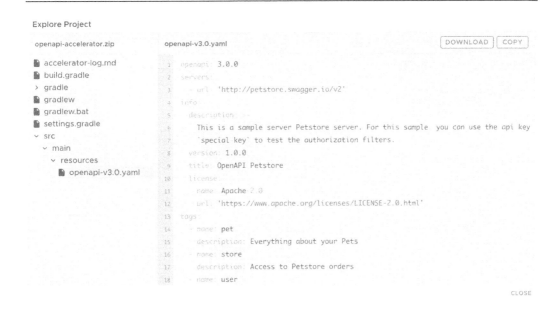

Figure 2.10 – Exploring our accelerator after plugging in a custom API spec

4. Click **Next Step**, **Create**, and **Download Zip File**.

5. Unzip the project and navigate to that directory.

6. Build the Spring Boot project and run the app:

```
$ chmod +x gradlew && ./gradlew openApiGenerate  # (this
will create a maven project in-place, use gradlew.bat on
Windows)

...

$ mvn spring-boot:run
```

7. Point your browser to http://localhost:8080:

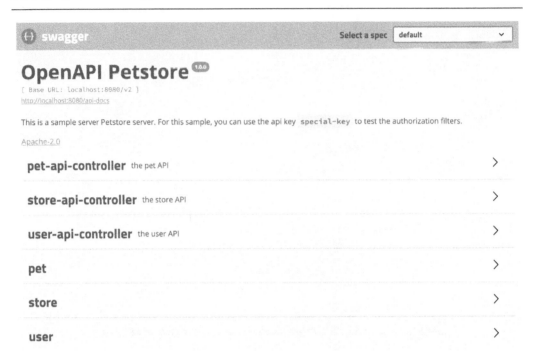

Figure 2.11 – Our custom OpenAPI 3.0 API in Swagger UI

And voila! With a simple download, we have a fully functional REST API and Swagger UI ready for us to start implementing business logic. We can start delivering value immediately because all our corporate governance, libraries, and opinions are conveniently baked into the project from its inception.

Congratulations! If you've made it this far, you've put pre-vetted, pre-audited, secure software frameworks in front of your developers, allowing them to skip over the tedious administrative work and get right down to delivering real value with their software.

However, the software is always changing and a software framework that was last week's up-to-date, rock-solid building block is this week's **Critical Vulnerability and Exposure (CVE)**.

Let's move on to some day-2 operations now and look at how we keep our app accelerators patched and up to date.

Day-2 operations with App Accelerator

In software development, getting something deployed into production is often the easy part. Keeping that software up to date and patched with minimal user disruption is where most of your time will be spent. App Accelerator was designed with day 2 in mind.

As it turns out, the component versions in our OpenAPI application accelerator have already gone out of date. Let's quickly update and patch our accelerator.

First, let's find our local copy of the `openapi-accelerator` project. Open `build.gradle` in your favorite editor. Notice the `org.springframework.boot` and `org.openapi.generator` dependencies are not the most recent and could therefore contain bugs or even known vulnerabilities. Notice that `org.springframework.boot` is at `2.5.6` and `org.openapi.generator` is at `5.2.1`:

```
38 lines (32 sloc)   869 Bytes

1   plugins {
2       id 'org.springframework.boot' version '2.5.6'
3       id 'io.spring.dependency-management' version '1.0.11.RELEASE'
4       id "org.openapi.generator" version "5.2.1"
5       id 'java'
6   }
7
```

Figure 2.12 – Our build.gradle file with out-of-date versions of Spring Boot and OpenAPI Generator

First, let's visit `https://start.spring.io` to find out what the latest release version of Spring Boot is. At the time of writing, it's 2.6.2:

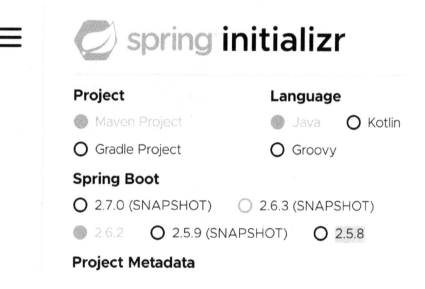

Figure 2.13 – Latest Spring Boot version from start.spring.io

We can also grab the latest version of the `openapi` plugin from `https://plugins.gradle.org/plugin/org.openapi.generator`. As you can see in the following screenshot, it's **5.3.1** at the time of writing:

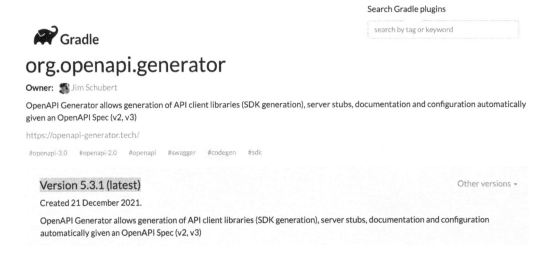

Search Gradle plugins

search by tag or keyword

🐘 **Gradle**

org.openapi.generator

Owner: 🧑 Jim Schubert

OpenAPI Generator allows generation of API client libraries (SDK generation), server stubs, documentation and configuration automatically given an OpenAPI Spec (v2, v3)

https://openapi-generator.tech/

#openapi-3.0   #openapi-2.0   #openapi   #swagger   #codegen   #sdk

Version 5.3.1 (latest)   Other versions ▾

Created 21 December 2021.

OpenAPI Generator allows generation of API client libraries (SDK generation), server stubs, documentation and configuration automatically given an OpenAPI Spec (v2, v3)

Figure 2.14 – The latest release version of the org.openapi.generator plugin (5.3.1)

Now that we have our up-to-date version numbers, let's update our `build.grade` file:

```
Users > roberthardt > work2 > tanzu > accelerator > openapi-accelerator > 🐘 build.gradle
 1  plugins {
 2      id 'org.springframework.boot' version '2.6.2'
 3      id 'io.spring.dependency-management' version '1.0.11.RELEASE'
 4      id "org.openapi.generator" version "5.3.1"
 5      id 'java'
 6  }
 7
 8  group = 'com.example'
 9  version = '0.0.1-SNAPSHOT'
10  sourceCompatibility = '11'
11
12  repositories {
13      mavenCentral()
14  }
15
16  dependencies {
17      implementation 'org.springframework.boot:spring-boot-starter-hateoas'
18      testImplementation 'org.springframework.boot:spring-boot-starter-test'
19  }
20
21  test {
22      useJUnitPlatform()
23  }
```

Figure 2.15 – Updating the plugins block in build.gradle with the latest versions

Then, we simply *git-push* our updated accelerator and after a few minutes, we can explore it in the UI and see our updated dependencies. Anyone who uses this accelerator in the future will get the latest versions of the `openapi` plugin and Spring Boot.

Now, it's time to switch gears. App Accelerator can help you get off to a great start in any number of popular development frameworks, but what are those frameworks? Are some better suited than others for the enterprise? The next section will cover two development frameworks especially well-suited to enterprise applications.

Cloud-native development frameworks under the Tanzu umbrella

As mentioned in previous sections and chapters, the Tanzu portfolio didn't materialize out of thin air. Rather, it is the thoughtful natural expression of many years of work and has helped some of the world's largest and most notable companies cultivate their software competency. One outcome of that experience is learning which development frameworks maximize developer productivity while minimizing toil, tedium, and bad practice.

At this point, we've successfully deployed, updated, downloaded, and implemented a simple API based on an application accelerator. Our example used Spring Boot. How did we arrive at Spring Boot as the underlying application framework for our API?

The Spring Framework and Spring Boot

The framework of choice for many of the most successful Tanzu customers has been the Spring Framework. This is proven by the millions of applications started with Spring Initializr at `https://start.spring.io`, and enterprise surveys such as the Snyk JVM Ecosystem Report 2020 showing fully 60% of the enterprise running Spring Boot (`https://snyk.io/blog/spring-dominates-the-java-ecosystem-with-60-using-it-for-their-main-applications/`).

Here's a screenshot showing some of the things Spring gives you right out of the box:

What Spring can do

Microservices
Quickly deliver production-grade features with independently evolvable microservices.

Reactive
Spring's asynchronous, nonblocking architecture means you can get more from your computing resources.

Cloud
Your code, any cloud—we've got you covered. Connect and scale your services, whatever your platform.

Web apps
Frameworks for fast, secure, and responsive web applications connected to any data store.

Serverless
The ultimate flexibility. Scale up on demand and scale to zero when there's no demand.

Event Driven
Integrate with your enterprise. React to business events. Act on your streaming data in realtime.

Batch
Automated tasks. Offline processing of data at a time to suit you.

Figure 2.16 – What Spring can do (spring.io)

The Spring Framework is very flexible in how it can be deployed. Spring Boot is an opinionated way to run Spring Framework applications so that they are self-contained and *just run*. Spring Boot gives a Spring Framework developer some very useful tools to run self-contained applications so that they're ideal for microservices. It also builds in sensible defaults and opinions that work for the most common scenarios. Spring Boot gives you the following:

- Several best-of-breed web frameworks to choose from

- Standalone runnable applications, perfect for containers and Kubernetes

- Opinionated *starters* for common use cases

- Autoconfiguration of all libraries wherever it's possible

- Real-world production features such as metrics and Kubernetes health checks

There are entire books written about the Spring Framework and Spring Boot, so we can't do it justice in this chapter. Instead, I'll give you some links for further exploration:

- Mark Heckler's *Spring Boot: Up and Running*: `https://www.oreilly.com/library/view/spring-boot-up/9781492076971/`

- Spring developer quickstarts: `https://spring.io/quickstart`
- Spring guides and tutorials: `https://spring.io/guides`
- Josh Long's *A Bootiful Podcast*: `https://soundcloud.com/a-bootiful-podcast`
- SpringDeveloper YouTube channel: `https://www.youtube.com/user/SpringSourceDev`

Perhaps you're asking yourself, *what does the open source Spring Framework have to do with Tanzu?* Excellent question! Among other things, many products in the Tanzu portfolio are enterprise versions of open source Spring projects. They include the following:

- Spring Cloud Gateway (`https://spring.io/projects/spring-cloud-gateway`)
- Spring Cloud Data Flow (`https://spring.io/projects/spring-cloud-dataflow`)
- Spring Cloud (`https://spring.io/projects/spring-cloud`)

There are also partnerships with cloud providers to bring an end-to-end, vertically integrated Spring platform such as Azure Spring Cloud (`https://tanzu.vmware.com/azure-spring-cloud`).

As amazing and rich as the Spring Framework is, it is geared toward Java developers. What if you're a developer with a background in the Microsoft technology stack? We'll discuss that next.

Steeltoe framework

Of course, not every enterprise software shop has career Java developers. Many millions of lines of code currently running the modern world are written on the Microsoft .NET stack. *Steeltoe* is a set of libraries, very much in the spirit of Spring Boot, that allow .NET developers to deliver cloud-native microservice applications reliably and efficiently. If you have an interest in maximizing the productivity of developers delivering enterprise software on .NET, I'd highly recommend you try it out:

`https://dotnetfoundation.org/projects/steeltoe`

And there you have it! With Spring and Project Steeltoe, we've covered application frameworks and toolsets that will appeal to the vast majority of enterprise developers, making them more productive, less prone to technical debt and bad decision-making, and most importantly, happier with their job, delivering economic value with software.

Summary

In the ongoing quest to maximize our productivity as developers or to maximize our teams' productivity as tech leaders or architects, we need to get *day 0* exactly right. Bad decisions or development mistakes made on day 0 will continue to bear their bitter, rotten fruit year after year.

The best way to avoid these costly day-0 mistakes is to begin from a starting point that we know has been successful in the past and use it as a jumping-off point for future development. Application Accelerator for VMware Tanzu, Spring Boot, and the Steeltoe framework provide a simple, repeatable baseline for codifying these starting points and making teams successful from the outset.

Furthermore, production-grade, enterprise-ready development frameworks don't appear overnight. Frameworks such as Spring and .NET + Steeltoe are battle-hardened based on decades of lessons learned. The best app accelerators use the best underlying frameworks and they're all part of the Tanzu portfolio.

Having worked through this chapter, you now have the skills to deploy and consume application accelerators that allow you to get day 0 exactly right – and furthermore, you're off to a great start with industry-standard application frameworks.

Now that you've solved the problem of getting out of the gate with a great application, let's learn about getting that application into a format that will run on a container platform such as Kubernetes next. Just as day-0 project inception is fraught with hazards, creating lean, secure, and repeatable container images presents another class of problems. Those problems are exactly what we'll cover in the next chapter when we discuss *Tanzu Build Service*.

Building Secure Container Images with Build Service

In the previous chapter, we discussed how Application Accelerator for VMware Tanzu helps organizations with a uniform and efficient way of building greenfield applications. This is a great start to building cloud-native applications that are based on predefined templates. These templates help developers purely focus on the business logic, which brings revenue to the organization.

> **Greenfield and cloud-native applications**
>
> Greenfield is a term from the construction industry that refers to undeveloped land. In the IT world, greenfield describes a software project that is developed from scratch rather than built from an existing program. It is often contrasted with *brownfield*, which describes software built from an existing program. Reference: `https://techterms.com/definition/greenfield`.
>
> Cloud-native applications, as you might surmise, are written to take advantage of cloud computing. They are characterized by such technologies as containers, service meshes, microservices, immutable infrastructure, and declarative APIs. Reference: `https://github.com/cncf/toc/blob/main/DEFINITION.md`.

However, to get the true benefit out of a cloud-native application on a container platform such as Kubernetes, we need to run these applications as containers. And, to run them as containers, we need to build container images for those applications. While there are various ways we can build such container images for our applications, one of the most popular approaches in the industry is to build them using configuration files known as Dockerfiles. A **Dockerfile** contains the definition, requirements, and attributes of the container image that should be built for the application. Though using Dockerfile is one of the most popular approaches to building container images, it is not always the most optimal one.

In this chapter, we will take a deep dive into this concept and cover the following topics:

- Why Tanzu Build Service?
- Unboxing Tanzu Build Service
- Getting started with Tanzu Build Service
- Common day-2 activities for Tanzu Build Service

So, let's get started.

Technical requirements

Some technical requirements need to be fulfilled before we start installing **Tanzu Build Service** (**TBS**). These requirements will be covered later in this chapter at the beginning of the *Getting started with Tanzu Build Service* section. However, you may not need them to understand the benefits TBS brings to the table. Let's start looking into it.

Why Tanzu Build Service?

There are various business, technical, and security challenges in building container images for applications. This becomes even more complex when we do it at scale in a large enterprise. Let's understand what those challenges are and how TBS addresses them.

Increasing developer productivity

As discussed, one of the most popular approaches to building container images today is using Dockerfiles. And, in most cases, the application teams are responsible for building and maintaining such Dockerfiles for their applications. These Dockerfiles contain details such as the base container operating system and its version, application bundles such as JAR files for a Java application, environment variables, and useful libraries and their versions.

> JAR files
>
> A **Java ARchive** (**JAR**) file is a package of an application containing compiled source code files, configuration files, and external libraries required by the application. A JAR file can either be a supporting library or an application package that can be run in a **Java Runtime Environment** (**JRE**).

Developers know their applications more than anybody. So, it makes sense that they define what goes in their applications' Dockerfiles. But at the same time, building and managing Dockerfiles are additional overheads for developers. Developers should spend all their time building more business-impacting functionalities in their applications. You might argue that building and changing such Dockerfiles is not a frequent task. Also, you may build some automation around building containers to reduce the amount of effort. However, such in-house automation brings other maintenance challenges. It would not eliminate the time required from the application teams. It's not just about the time the developers need to spend to create or update the Dockerfiles. They also need the time required to research and decide on the content of it. And finally, these Dockerfiles have to be kept up to date to reflect the latest security patches of the libraries referenced in them. That ensures the best possible security posture for the running containers. Such endless ongoing maintenance consumes a lot of productive time of developers for unproductive activities.

Layers in container images

A final container image of an application could be a combination of multiple smaller images that are stacked as layers on top of each other to provide reusability, separation, and ease of usage.

To address these challenges, Pivotal Software Inc. (which was acquired by VMware Inc. in 2020) and Heroku collaborated. They incepted an open source project called **buildpacks.io** under the **Cloud Native Computing Foundation** (**CNCF**). We will discuss this project later in this chapter in detail. TBS is commercially-supported packaging containing buildpacks.io and a few other open source tools.

TBS addresses this challenge by providing a complete automation engine to build container images when you supply application code or built artifacts. As an output, TBS generates an OCI-compliant container image for the application. This image can be deployed on Kubernetes or any other OCI-compliant container orchestration platform. With TBS, developers are off the hook to build and maintain their container images. The reduced amount of responsibilities helps developers focus on what is more important for the business.

What is OCI?

Open Container Initiative (**OCI**) is a standard set by The Linux Foundation describing the characteristics of a container image that can be implemented by various container image-building tools and understood by different container scheduling platforms such as Kubernetes. All major container platforms, including Kubernetes, support OCI-compliant container images.

TBS supports different languages including Java, .Net, Python, Go, NodeJS, and many more.

Reduction of bespoke automation

It is a commonly observed practice that organizations create **continuous integration** (**CI**) pipelines to build their applications' container images. These pipelines are often developed using tools such as Jenkins and written mostly using languages such as Python or Shell script. Organizations may need to invest in resources to first develop such custom automation and then to maintain them ongoingly. Furthermore, the lack of good documentation around such custom automation makes such maintenance a nightmare. Hence, such people-dependent automation becomes a pain to maintain when their parents leave the organization. Additionally, the organizations could get better business outcomes if these people could rather be used for a better business-value-oriented assignment instead of such below-value-line engineering efforts.

TBS also helps address this challenge. It automates the container image-building process to a significant level. Although this will not replace the entire CI pipeline, it will reduce its complexity by covering the various steps required to build container images with full automation.

Standardization of container build process

It is often seen that there are many departmental silos in enterprises with large development shops. Such silos have tools and practices to follow. This could be a huge waste of crucial resources for the organizations in terms of duplication at various levels. Such duplication could be people's time spent for similar outcomes, the license cost of tools, and the infrastructure used by the automation. This could lead to a whole new issue of lack of standardization. Such non-standard practices result in decreased transparency, governance, and security posture. When it comes to building container images, such an absence of standardization could be proven to be a very costly mistake for security risk exposures. This is caused by using unapproved libraries or not patching them quickly. When different teams have different ways of building container images, they could follow different practices. They might use different container operating systems, open source tools, third-party libraries, and their versions. It would be very difficult to apply an enterprise-wide standard. Such enforced standards should not affect different teams' productivity and freedom of choice.

TBS solves this problem in two ways. First, it includes a centralized software library provided by VMware in the form of buildpacks and Stacks. Here, **buildpacks** include all the required libraries for the application to work in the container, including application runtimes such as **Java Runtime Environment** (JRE) and middleware such as Tomcat Server. On the other side, **Stacks** include different flavors of container operating systems. Second, TBS provides a standardized container image-building automation engine. Hence, when an organization uses TBS to build containers, it automatically implements standardization in the container-building process. This standardization comes in the form of the required automation and the application supporting content in the images.

TBS does not only help to standardize the container build process across the company but also improves the overall security posture around the same, as explained in the following section.

Stronger security posture

Security exposure is a major concern for the most established organizations in their journey to cloud transformation. Most cloud-native applications are deployed as containers in either public or private cloud platforms for several benefits. Containers are nothing but tiny virtual machines where the applications run. It is critical that such containers are built with secure ingredients that do not contain security vulnerabilities. But today's secure library version could be found vulnerable tomorrow since it is very common to see new security vulnerabilities getting announced often for all operating systems and libraries that could be used in those container images. The corresponding organizations behind such operating systems and libraries would release newer versions to address those **Common Vulnerabilities and Exposures** (**CVE**). However, it is on the user organizations to take the newer versions of the software and use them to rebuild the impacted container images. Such container image rebuild exercises may introduce two big pain points. The first is when there are multiple development teams managing hundreds of containerized applications. In that case, it is very difficult to find the impacted applications. Such an identification and remediation process may take weeks, keeping those applications vulnerable to attacks. It would be very difficult to push all the application teams at the same time to rebuild their applications' container images using the newer version of the software. The patching gets delayed as the application teams would have their product backlogs and priorities to manage. And history has proven time and again that most major software-related security breaches were driven by unpatched software components running for a long time.

TBS can greatly speed up the CVE patching of the impacted container images with its centralized resource library and container image rebuild automation using that library. As a new CVE patch is announced for the impacted component in the library, VMware releases a new version of the repository component (either a buildpack or a stack) containing the fix for the vulnerability. TBS identifies impacted application container images using an internal map of container images and their linked dependencies. So, when VMware releases a new component version in the centralized repository to fix any CVE, TBS immediately triggers the patching of the impacted application images using the patched version. Such an automatic rebuilding of images for hundreds of applications may be complete in a few hours rather than needing weeks in the absence of TBS. This could be one of the most important reasons to consider using a tool such as TBS. Such mappings of applications, their container images, and the associated software components used in them provide the required transparency and auditability to the security team. Using TBS, we can quickly generate a **Bill of Material** (**BOM**) for any application managed by TBS. A BOM is a detailed report listing all the components and their versions used in the respective container images. It greatly simplifies security audits of containerized environments.

Optimized network bandwidth and storage utilization

As described previously, an OCI-compliant container image is a collection of other smaller images containing different components required for the final resultant container image. These layers are made up of application code, configurations, third-party libraries, and operating systems. Container image repositories such as Docker Hub and Harbor store such layers separately. They also maintain maps of which image layer depends on which other image layers. So, when you pull a specific container image from the container image registry, the registry pushes all dependent container image layers as a result. Due to this, when an application goes through any change, only the corresponding impacted layer will get transferred over the network into the container registry. All other non-impacted image layers will not move over the network. This makes the image push and pull operations a lot more efficient. It also helps to reduce the storage requirements for the container image registry because of the reuse of the unchanged layers. However, to get the full benefit of these layers, you should follow some discipline in the image-building process. If the authors of Dockerfiles do not take enough care, they end up with fewer image layers than what is optimally possible. The following figures show the anatomy of the same application's container images. They both have a different number of layers. They are built with two different Dockerfile approaches. These snapshots were taken using an open source tool named **Dive** (`https://github.com/wagoodman/dive`), which gives a detailed view of layers and their content for a container image:

```
| • Layers |
Cmp   Size   Command
      73 MB   FROM 3f26145e37da197
      43 MB   apt-get update      && apt-get install -y --no-install-recommends tzdata curl ca-certificates fontconfig lo
     128 MB   set -eux;      ARCH="$(dpkg --print-architecture)";     case "${ARCH}" in       aarch64|arm64)        ES
      48 MB   COPY target/*.jar app.jar # buildkit
```

Figure 3.1 – A demo Java application's container image built using four layers

As you can see, *Figure 3.1* has only four layers, whereas *Figure 3.2* has eight layers, even though the total resultant image size is almost the same. There are various ways to build a container image, which may result in different outputs for the same application code:

```
| • Layers |
Cmp   Size   Command
      73 MB   FROM 3f26145e37da197
      43 MB   apt-get update      && apt-get install -y --no-install-recommends tzdata curl ca-certificates fontconfig lo
     128 MB   set -eux;      ARCH="$(dpkg --print-architecture)";     case "${ARCH}" in       aarch64|arm64)        ES
       0 B   WORKDIR /application
      48 MB   COPY application/dependencies/ ./ # buildkit
     246 kB   COPY application/spring-boot-loader/ ./ # buildkit
       0 B   COPY application/snapshot-dependencies/ ./ # buildkit
     102 kB   COPY application/application/ ./ # buildkit
```

Figure 3.2 – The same demo Java application's container image but built using eight layers

When the developers do not have the required awareness or there are no enterprise-level guidelines on how to build application container images, every team might end up with their own standards and practices. With such a lack of knowledge and controls in place, large organizations may end up with several suboptimal application container images, which may lead to a waste of network bandwidth and storage for every container image push and pull operation.

On the other side, TBS uses a highly acclaimed **Cloud Native Computing Foundation** (**CNCF**) (`https://cncf.io`) certified project named **Cloud Native Buildpacks** (**CNB**) (`https://buildpack.io`) as a tool under the hood, which provides a way to build container images with several smaller layers when the application container images are built using this tool. It provides an organization-level standardized approach to building container images that are also very resource efficient, along with having other benefits, as discussed previously. Here is a high-level representation depicting how TBS performs this operation:

Figure 3.3 – High-level representation of how TBS builds container images with layers

The preceding figure shows how TBS takes some application code, performs various operations internally, and creates a final OCI-compliant application container image that has different smaller layers.

Overall, TBS is useful to enhance developer productivity, reduce the learning curve, reduce operational toil, and increase security posture, along with several other benefits listed here for building secure container images. With all that, it just helps you accelerate your cloud-native application journey. After learning about the different benefits of using a tool such as TBS, let's unbox it to check its anatomy. We will take a deep dive into all the different components that are bundled together as TBS.

Unboxing Tanzu Build Service

As described previously, TBS is built on top of two main open source projects: CNB and kpack. The following figure depicts the whole packaging of TBS:

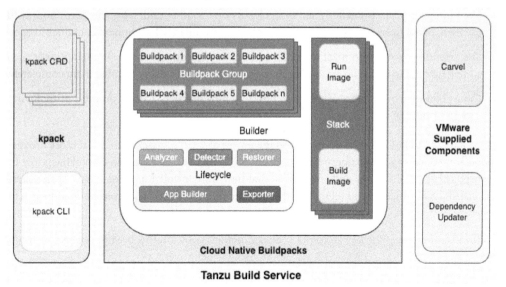

Figure 3.4 – Building blocks of Tanzu Build Service

As shown in the preceding figure, kpack includes engine and CLIs, whereas CNB includes things such as Builder, buildpacks and their groups, the stack, build and run images, and the life cycle process. Additionally, there are some VMware-supplied components for additional functionalities that are bundled in TBS. Let's understand them in detail.

Cloud-native buildpacks

The concept of CNB was derived from the concept of buildpacks in Cloud Foundry, which is another container orchestration platform for cloud-native applications. The buildpacks in Cloud Foundry have been proven a battle-tested tool for over a decade. Buildpacks in Cloud Foundry are used to scan application source code, determine application requirements based on the technology and language used, club all required dependencies with application packages, and create an offline container called a **droplet**. These droplets are large binary objects that contain everything required by the corresponding applications to run as containers on the Cloud Foundry platform. Though droplets have a solid track record, they have some limitations, listed as follows:

- Cloud Foundry buildpacks generate droplets that are very large as they contain everything required by the application. They do not have a layer concept like modern OCI-compliant container images do. Hence, every small change in the application or its dependency creates a new version of the heavy droplet again. Such droplets require more storage as they do not contain just the delta but a full-blown application package. Because of this, deploying containers using droplets in Cloud Foundry is a relatively slower process than deploying containers using an OCI image for a newer version of the application on an OCI-compatible platform such as Kubernetes. This slowness of container scheduling impacts application deployment times while scaling or redeploying the applications.

- The Buildpack project was designed to work only on Cloud Foundry. So, there was no way to use them on platforms such as Kubernetes that support only OCI-compliant container images.

- The extendibility or customization of Cloud Foundry buildpacks is very limited. Rather than adding newer changes as layers or plugins, they must be modified by opening them completely.

- Cloud Foundry buildpacks are slow in building new droplets as they do not optimally use cached resources.

Keeping all these limitations in mind and using the power of the buildpack concept in more broader and popular platforms such as Kubernetes, two software companies, Heroku and Pivotal (acquired by VMware in 2019), joined hands and announced that they were to collaborate on a new open source project called CNB, which retains all the goodness of the original Buildpack project but addresses its weaknesses. With this background, let's discuss the anatomy of CNB in detail.

The following are some of the key concepts and terminologies of CNB that are important constructs of TBS.

Build image

This is a base container operating system layer that is used to create a builder's container image. A build container is short-lived. It executes the life cycle process to build the application's container image and gets terminated.

Run image

This is a base container operating system layer that is used to create an application's container image. This is the main outcome of TBS. All the application-specific container images built by TBS use this image as the base container operating system layer.

Stack

This is a configuration entity that contains details of the build image and runs image flavors to be used in the container image build process. As a part of the package, TBS provides four different flavors of stacks that contain different flavors of build-and-run operating systems. These operating systems layers could be either thin or thick, depending on the requirements of the applications that need to use them.

Buildpack

This is a collection of executables that inspect your application code and determines whether the buildpack would apply to the application and hence should be a part of the resultant application's container image. For example, there is a buildpack for Java applications, which detects the presence of Java-specific files in the application code or artifact, and then takes a call if the application needs required support to run as a Java application. In the TBS architecture, such buildpacks are stored in a container registry as container images.

Buildpack group

This is a collection of buildpacks that are typically used together to build a container image of a specific type of application. The buildpacks that are members of a buildpack group could either be mandatory or optional, depending on their use case and the requirement of the application. For example, buildpacks for **Java Runtime Environment** (**JRE**), Maven, Gradle, Tomcat Server, and Jetty Server may all fall in the same group as they are all Java application-related dependencies. However, the buildpack for JRE would be a compulsory one for a Java application, but all others listed before would fall into the optional category as the application might or might not need them. There are various such buildpack groups for different types of applications, such as Python, .Net, NodeJS, and more.

Life cycle

This determines the application of buildpacks and orchestrates their execution. The life cycle process has various components that execute its stages. At the end of all the stages, we get the final OCI-compliant container image of an application. Let's review these life cycle components:

- **Analyzer**: This retrieves and examines all required files that would be required during the image build process and used by the buildpacks. It also checks whether all the required images for the build are accessible in the container registry used by TBS. With such quick checking of dependencies, the build process fails fast if something is missing rather than realizing that later in the build process.

- **Detector**: This checks which buildpack group is applicable for the application under the build process. It also gets a list of all available buildpack groups in a specific order. Then, it checks the applicability of each of these groups one by one until the first group passes the required criteria. The detector then creates a plan of execution to perform the container image build process as a resulting artifact.

- **Restorer**: This restores all the required dependency image layers based on the selected buildpack group from the cache put in by previous container build processes using the same image layers. This phase reduces the build time and the network traffic to transfer frequently used images.

- **App Builder**: This transforms the application source code into a runnable artifact that can be packaged for execution inside a container. For example, this stage converts a Java application source code into a JAR file with compiled class files as an executable artifact. The application of this stage could be optional, depending on the supplied artifact of the application or even based on the technology requirements. For example, if the build process gets an already prepared JAR file for a Java application instead of the source code, then there is no need to prepare a build for the application and this stage may be skipped.

- **Exporter**: This creates the final OCI-compliant container image file. It also prepares a report containing the BOM for the components and their versions used in the container image. Finally, it pushes the container image into the target registry.

Builder

To build a container image of an application, TBS needs to deploy a temporary container on Kubernetes that executes the life cycle components, as previously described, to create the resultant container image. A builder is a container image that deploys this container, which contains the executables for the life cycle processes, along with a list of buildpack groups and a build image.

kpack

kpack (`https://github.com/pivotal/kpack`) is an open source project initiated by Pivotal and is now actively being maintained by VMware. kpack provides a way to use CNB on the Kubernetes platform. kpack uses some Kubernetes **Custom Resource Definitions** (**CRDs**) to deploy itself as a tool running on top of Kubernetes. Hence, kpack is a tool that runs on Kubernetes and uses CNB to build OCI-compliant container images.

Custom Resource Definitions (CRDs)

Kubernetes has several out-of-the-box APIs that are referred to as resources. Some examples of such resources are Pod, Node, Deployment, Service, ReplicaSet, and many others. Though Kubernetes comes with many such out-of-the-box resources, it is a very extensible platform that allows adding more resources that are custom. Such custom resources are called CRDs. You may learn more about CRDs here: `https://kubernetes.io/docs/concepts/extend-kubernetes/api-extension/custom-resources/`.

---

kpack has two main components, as depicted in *Figure 3.4*:

- kpack Kubernetes CRDs to help use CNB and define the container image specification.
- The kpack **command-line interface** (**CLI**), which provides the required user interface to use kpack resources. The kpack CLI provides ways to create and manage container image build specifications using kpack.

kpack CRD objects used by TBS

Though kpack is an internal component of TBS, the following are some of the key kpack CRD objects that are used by TBS:

- **Image**: An image gets created for every application that is registered with TBS to build its container images. An `Image` object references the application source code or package location, the runtime details for the build process, and the container registry details to store the built images.
- **ClusterStore**: This contains references to buildpacks in the form of their respective container image locations.
- **ClusterStack**: This contains references to the OS layers in terms of the build image and run image, along with their container registry locations.
- **ClusterBuilder**: This is a high-level map that links `ClusterStore` and `ClusterStack` combinations. A `ClusterBuilder` object also defines an order of buildpacks to be validated against any application.

VMware-provided components and features

In addition to the two main open source build blocks of TBS – CNB and kpack – there are also a few additional components and functionalities provided by VMware that the enterprises can get as a part of TBS packaging. Let's quickly visit them:

- TBS comes with a proprietary installation and upgrades the user experience using an open source toolkit named Carvel (`https://carvel.dev`). It is a Kubernetes application packaging and deployment toolkit mainly maintained by VMware. We will use it to install TBS in the next section.

- TBS has a dependency updater component that keeps all the container images built by TBS up to date with changes made in their corresponding buildpacks or stack. This feature helps to keep all the container images patched and secured with the latest updates in the operating system and application dependency changes.

- TBS also comes with a bundle of VMware-supplied buildpacks. This includes the support for offline buildpacks, Windows containers, and quick and reliable release engineering of the new buildpack versions to include new features and fixes of CVEs.

In this section, we saw the structure and components of TBS. We also learned the role that each of them plays to build the whole solution. Now, let's get started with working with TBS. In the next section, you will learn how to install and configure TBS in your Kubernetes cluster and rip all the benefits we discussed previously.

Getting started with Tanzu Build Service

After learning about the challenges addressed by TBS and the details of what TBS contains, let's learn how we can quickly get started with it running in a Kubernetes cluster.

> **Important note**
>
> All these instructions are for **Tanzu Build Service (TBS)** v1.3.

The following section details different prerequisites that you may need to get TBS fully up and running in your Kubernetes cluster.

Prerequisites

You will need the following to configure TBS in your Kubernetes environment:

- Administrator-level `kubectl` CLI access to a Kubernetes cluster with version 1.19 or later. If administrator-level access is not feasible, then the user must at least have permissions listed at `https://github.com/tandcruz/DevSecOps-in-Practice-with-VMware-Tanzu/blob/main/chapter-03/tbs-k8s-cluster-permissions.yml` on the cluster to install and configure TBS.

- The worker nodes of the Kubernetes cluster should at least have 50 GB of ephemeral storage as TBS stores the historical versions of the built images for records. The number of historical versions stored by TBS can be configured. This will be covered later in this chapter under day-2 activities.

- Access to any container registry with required permission that supports Docker HTTP API V2 with at least a 5 GB storage quota, which excludes the space required for application images built by TBS. To keep things simple, we will use Docker Hub (`https://www.dockerhub.com/`), which provides a free account that is good enough for TBS integration.

- There should be a default **StorageClass** configured in your Kubernetes cluster that TBS can use to create the required storage volumes. By default, TBS will need a **PersistentVolumeClaim** that it uses to cache already-built artifacts. Such caching helps the subsequent builds complete faster.

- The operator machine that will be used for this installation should have Carvel CLI tools installed. The following are those Carvel tools that TBS uses, along with their download locations and purposes:

  - kapp version 0.41.0 (`https://network.tanzu.vmware.com/products/kapp/`) to deploy the bundle of Kubernetes resources required for TBS.

  - ytt version 0.37.0 (`https://network.tanzu.vmware.com/products/ytt/`) to replace custom configuration values in the YAML template files used for TBS Kubernetes resource deployments.

  - kbld version 0.31.0 (`https://network.tanzu.vmware.com/products/kbld/`) to reference container images in Kubernetes configuration files that are relocated based on your choice of container registry.

  - imgpkg version 0.23.1 (`https://network.tanzu.vmware.com/products/imgpkg/`) to deploy the packaged application bundle for TBS that contains the required configuration and OCI images. For an air-gapped (an environment that has no outbound internet connectivity) installation, it helps to relocate all the required OCI images to the private container registry in use.

- The operator machine should have kp CLI v0.4.*, which can be downloaded from the Tanzu Network website at `https://network.tanzu.vmware.com/products/build-service/`.

- The operator machine should have the docker CLI: `https://docs.docker.com/get-docker/`.

- The operator machine should have the Dependency Descriptor file in the `descriptor-<version>.yaml` format downloaded from the TBS dependencies page on the Tanzu Network website at `https://network.tanzu.vmware.com/products/build-service/`. This book has used the `descriptor-100.0.229.yaml` file. This file contains container image paths that TBS will need to execute image builds. You may find a different version, depending on when you download it, which is fine.

Important note

The Kubernetes cluster running with Containerd v1.4.1 is not compatible with TBS. The following `kubectl` command will get the version of the underneath container runtime to check this:

```
kubectl get nodes -o=jsonpath='{.items[0].status.nodeInfo.containerRuntimeVersion}'
```

Let's start our journey working with TBS by first installing it and then performing some basic tests to confirm whether it is working as expected.

Installation procedure

In this chapter, we will use Docker Hub as the container registry to be used with TBS. Also, the installation steps assume that the base Kubernetes cluster has full outbound internet connectivity. The procedure to install and configure TBS is different for an air-gapped environment and a custom container registry to be used instead of Docker Hub. You may follow the official product documentation (https://docs.vmware.com/en/Tanzu-Build-Service/1.3/vmware-tanzu-build-service-v13/GUID-installing.html) for a different use case.

Additionally, the p********t value used in all the commands should be replaced with your respective username and ********** with your respective password.

With those expectations set, let's install and configure TBS by performing the following steps:

1. Make sure you are working in the right Kubernetes cluster and context where you want to install TBS:

   ```
   $ kubectl config use-context <CONTEXT-NAME>
   ```

2. Relocate the required container images from the Tanzu Network registry to your Docker Hub account. For that, log in to your Docker Hub account, as follows:

   ```
   $ docker login -u p********t
   Password:
   Login Succeeded
   ```

3. Log in to the Tanzu Network container registry to pull the required images for installation using your Tanzu Network credentials:

   ```
   $ docker login registry.tanzu.vmware.com
   Username: p********t@*******.io
   Password:
   Login Succeeded
   ```

4. Relocate the images from Tanzu Network to your Docker Hub registry using the following
    `imgpkg` command:

    ```
    $ imgpkg copy -b "registry.tanzu.vmware.com/build-
    service/bundle:1.3.0" --to-repo p********t/build-
    service-1-3

    copy | exporting 17 images...
    copy | will export registry.tanzu.vmware.com/build-ser-
    vice/bundle@sha256:0e64239d34119c1b8140d457a2380507513606
    17d8e8b64703d8b7b4f944054a
    ..
    copy | will export registry.tanzu.vmware.com/build-ser-
    vice/smart-warmer@sha256:4d865b7f4c10c1099ae9648a64e6e7d-
    a097d0a375551e8fd2ef80a6d1fc50176
    copy | exported 17 images
    copy | importing 17 images...
     443.39 MiB / 443.64 MiB [===============================
    ===========================================================
    ======]  99.94% 1.92 MiB/s 3m51s
    copy | done uploading images
    copy | Warning: Skipped layer due to it being
    non-distributable. If you would like to include
    non-distributable layers, use the --include-non-
    distributable-layers flag
    Succeeded
    ```

    You may ignore the warning given before the success message because TBS excludes Windows
    components by default as they are licensed.

5. Pull the TBS bundle image locally using the `imgpkg` command:

    ```
    $ imgpkg pull -b "p********t/build-service-1-3:1.3.0" -o
    /tmp/bundle

    Pulling bundle 'index.docker.io/p********t/build-ser-
    vice-1-3@sha256:0e64239d34119c1b8140d457a238050751360617d
    8e8b64703d8b7b4f944054a'
      Extracting layer 'sha256:872d56ff2b8ef97689ecaa-
    0901199d84e7f7ae55bfef3ad9c7effa14b02e6dfd' (1/1)
    Locating image lock file images...
    ```

The bundle repo (index.docker.io/p********t/build-ser-vice-1-3) is hosting every image specified in the bundle's Images Lock file (.imgpkg/images.yml)

Succeeded

6. Install TBS using the relevant Carvel tools – ytt, kbld, and kapp – with the following command. It is a very long command that injects the provided custom parameter values with the -v flag into the deployment configuration files using ytt. Then, the command replaces the container image locations based on your registry location using kbld. And finally, it deploys TBS using the configuration files with custom parameter values and the required container image files pulled from your repository using kapp:

```
$ ytt -f /tmp/bundle/values.yaml -f /tmp/bundle/
config/ -v kp_default_repository='p********t/build-ser-
vice-1-3' -v kp_default_repository_username='p********t'
-v kp_default_repository_password='**********' -v
pull_from_kp_default_repo=true -v tanzunet_user-
name='p********t@*******.io' -v tanzunet_pass-
word='**********' | kbld -f /tmp/bundle/.imgpkg/images.
yml -f- | kapp deploy -a tanzu-build-service -f- -y

resolve | final: build-init -> index.docker.
io/p********t/build-service-1-3@sha256:838e8f1ad7be81e8d-
ab637259882f9c4daea70c42771264f96be4b57303d85f2
resolve | final: completion -> index.docker.
io/p********t/build-service-1-3@sha256:765dafb0bb1503ef-
2f9d2deb33b476b14c85023e5952f1eeb46a983feca595c6
. .

. .

Succeeded
```

> **Important note**
>
> At the time of writing this book, all the binaries supplied under Carvel, including ytt, kbld, and kapp, are unsigned binaries. Because of this, your operating system, especially macOS, may raise a security concern against using them. However, you may explicitly allow the execution of these binaries in your operating system's security settings. Additionally, as this command performs various long image pull operations to deploy TBS in your Kubernetes cluster, you may see the command complete unsuccessfully with an error – use of closed network connection. In that case, you may run the same command again and it may just work.

You may need to replace the highlighted values in the preceding command as per the following specification:

- Replace p********t in the kp_default_repository and kp_default_repository_ username parameters with your Docker Hub username
- Replace ********** in the kp_default_repository_password parameter with your Docker Hub account password
- Replace p********t@*******.io in tanzunet_username with your Tanzu Network username
- Replace ********** in tanzunet_password with your Tanzu Network password

With that last command completed successfully, TBS should be up and running in your Kubernetes cluster. Let's verify the installation and ensure TBS is working fine.

Verifying the installation

To verify the TBS installation, execute the following kp command to list the cluster builders available in your TBS environment:

```
$ kp clusterbuilder list
```

The result of the preceding command should look as follows, where you should see your Docker Hub username instead of p********t:

```
NAME        READY       STACK                           IMAGE
base        true        io.buildpacks.stacks.bionic p********t/
build-service-1-3:clusterbuilder-base@sha256:7af47645c47b-
305caa1b14b3900dbca206025e30b684e9cd32b6d27f9942661f
default     true        io.buildpacks.stacks.bionic     p********t/
build-service-1-3:clusterbuilder-default@sha256:7af47645c47b-
305caa1b14b3900dbca206025e30b684e9cd32b6d27f9942661f
full        true        io.buildpacks.stacks.bionic     p********t/
build-service-1-3:clusterbuilder-full@sha256:714990fdf5e-
90039024bceafd5f499830235f1b5f51477a3434f3b297646b3d0
tiny        true        io.paketo.stacks.tiny           p********t/
build-service-1-3:clusterbuilder-tiny@sha256:29d03b1d4f45ce6e-
7947ab2bf862023f47d5a6c84e634727900a1625e661ee3b
```

If you see the preceding output, then congratulations to you as you have TBS running in your Kubernetes environment, waiting to build container images of your application!

Now that we've got started with TBS, let's investigate common day-2 operations that we can perform on TBS for various use cases.

Common day-2 activities for Tanzu Build Service

In this section, we will go through some useful operations we can perform on TBS.

Building application container images

In this section, we will learn how to register our application with TBS for the first time, create the first container image, run that container image locally, retrigger the image build process again by modifying the application configuration, and, finally, verify the newly created container image to reflect the application change. This will be an exciting journey to use TBS for its main purpose.

Registering an application with TBS

The main reason to use TBS is to gain the ability to build application container images in a fully automatic and secure way. Let's see how we can build container images of a cloud-native application using the TBS setup we have completed. We will use a sample Spring Framework-based application, Spring Pet Clinic, available at `https://github.com/tandcruz/DevSecOps-in-Practice-with-VMware-Tanzu/tree/main/spring-petclinic`. To follow along, you need to fork this project in your Git repository.

We will use the kp CLI to register our application with TBS. To register an application, we must create an `image` resource, which is a kpack CRD object to create a record of the application in its list. Once an image resource has been created, TBS creates a `build`, which is also a kpack CRD object that creates a container image of a registered application. There can be one-to-many relationships between an `image` and its `build` objects, depending on the number of instances to create a new container image for an application. But before we register our application, let's verify the current image objects:

```
$ kp image list
Error: no image resources found
```

As you might have guessed, we don't have any existing `image` objects managed by our newly deployed TBS in our Kubernetes cluster.

Additionally, we will also need to provide TBS with the credentials to our Docker Hub account so that it can push built images there. You may also use a different container registry or a Docker Hub account to push built application images. But, to keep things simple, we will use the same Docker Hub account that we used previously to install TBS. To supply the login credentials of our Docker Hub account, we need to create a Kubernetes `Secret` object, as follows:

```
$ DOCKER_PASSWORD="**********" kp secret create tbs-dockerhub-
cred -dockerhub p********t


Secret "tbs-dockerhub-cred" created
```

Now, let's work on creating an `image` object.

There are three different ways in which we may configure an application to use TBS for building container images. Let's take a look at them so that you can understand which one you should use when:

1. **Using a Git repository URL**: In this approach, we register the Git repository URL and the branch of the repository that we want to monitor for changes and trigger TBS image builds based on the changes committed in this branch. This is the most automated approach to creating container images as soon as application changes are merged in the final code branch. We will use this approach in this book.

2. **Using a local path**: In this approach, we supply either the location of the application's package such as a JAR file for a Java application that is precompiled and packaged on the local system, or provide the location of the application's source code on the local system. This approach is not fully automated using TBS and assumes that you have an external CI process that will explicitly call TBS whenever there is a need to create a container image of the application, rather than creating new images automatically based on the new changes pushed into the Git repository branch.

3. **Using a BLOB store location**: Like the local path approach, in this approach, we supply the BLOB store location, such as an S3 bucket in AWS, containing application source code packaged as either a `zip`, `tar.gz`, `tar`, or `jar` file. This approach is also used for explicit build triggers like the previous one.

Now, let's register the application, Spring Pet Clinic, to be used with TBS, along with its Git repository. See the following command and its results, which explain how to do this. The command uses the application's Git repository and the branch that we want TBS to monitor for changes and build container images from:

```
$ kp image create tbs-spring-petclinic --tag p********t/
tbs-spring-petclinic:1.0 --wait --git  https://github.com/
PacktPublishing/DevSecOps-in-Practice-with-VMware-Tanzu/tree/
main/spring-petclinic --git-revision main
```

In this command, replace the value of `--git` and `--git-revision` with the repository details that you would have forked.

> **Important note**
>
> This command may take several minutes to run since it's the first build and application registration. Also, it assumes that the Git repository is publicly accessible. But if the repository requires user credentials to pull the source code, you may need to create a TBS `secret` object for the Git repository credentials, as described here: `https://docs.vmware.com/en/Tanzu-Build-Service/1.3/vmware-tanzu-build-service-v13/GUID-managing-secrets.html#create-a-git-ssh-secret`.

As you can see, this command performs several operations, including accessing the application's source code, downloading all required dependencies, performing all CNB life cycle stages, and finally, pushing the application image into your Docker Hub registry, as specified in the command.

Let's check the presence of `image` objects for our application:

```
$ kp image list
NAME                         READY      LATEST REASON      LATEST
IMAGE                                                      NAME-
SPACE
tbs-spring-petclinic         True       CONFIG             index.docker.
io/p********t/tbs-spring-petclinic@sha256:45688e54b22ee96e798f-
3f28e09a81020acc69fa0db806690aeb2ba07ae3ab00      default
```

Here, we can see that one `image` object has been created called `tbs-spring-petclinic`. Now, let's check the number of `build` objects that have been created for our application:

```
$ kp build list
BUILD     STATUS      BUILT IMAG
E                                                         REASO
N     IMAGE RESOURCE
1         SUCCESS     index.docker.io/p********t/tbs-spring-pet-
clinic@sha256:45688e54b22ee96e798f3f28e09a81020acc69fa0d-
b806690aeb2ba07ae3ab00      CONFIG     tbs-spring-petclinic
```

Here, we can see one `build` object created for our application `image` object. We may see more `build` objects if there were a greater number of image builds triggered for application changes. The column named REASON indicates the reason to get this build triggered by TBS. There are the following possible reasons:

- **CONFIG** to indicate when a change is made to commit, branch, Git repository, or build fields on the image's configuration file

- **COMMIT** to indicate a build as a result of a change pushed in an application's code repository under TBS's watch

- **STACK** to indicate a change in the run image OS layer

- **BUILDPACK** to indicate a change in the buildpack versions that are made available through an updated builder

- **TRIGGER** to indicate a build triggered manually

Now that the application has been registered with TBS, when you commit a small change in the monitored branch, you should see a new build getting created in a few seconds, as shown here:

```
$ kp build list
BUILD      STATUS      BUILT IMAG
E
REASON     IMAGE RESOURCE
1          SUCCESS      index.docker.io/p********t/tbs-spring-pet-
clinic@sha256:45688e54b22ee96e798f3f28e09a81020acc69fa0d-
b806690aeb2ba07ae3ab00     CONFIG    tbs-spring-petclinic
2          BUILD-
ING                                COMMIT    tbs-spring-pet-
clinic
```

You may now pull the old and the new images to deploy their containers within your local Docker environment to verify the changes that have been made to the application. Now that we've learned how to create new builds of the container images for the registered applications, let's learn how to check the build logs to see the execution details of the life cycle stages.

Checking image build logs

To check the TBS logs of the newly built image, use the following command:

```
$ kp build logs tbs-spring-petclinic
..

..
Saving p********t/tbs-spring-petclinic:1.0...
*** Images (sha256:322010cd44fa9dc3bcb0bfadc7ba6873fb65f1b-
fe4f0bbe6cf6dc9d4e3112e84):
        p********t/tbs-spring-petclinic:1.0
        index.docker.io/p********t/tbs-spring-petclin-
ic:1.0-b2.20211213.072737
===> COMPLETION
Build successful
```

The output has been truncated for brevity.

With that, we've learned how to create the first application configuration, trigger a new build, and check build logs on TBS. Now, let's discuss another very important activity around TBS, which is to keep our container images always secured and patched with the latest versions of the software libraries and operating system layers used in the application container images.

Upgrading buildpacks and stacks

As we saw in the *Why Tanzu Build Service* section, one of the main benefits of using this tool is to enhance the security posture by staying up to date using the latest patched application dependencies and the OS layer used to build container images. As we know, buildpacks contain references to different software library versions and stacks contain the OS layers for the container image building. So, when there are new patch releases of the libraries that are referenced in the buildpacks or the OS in the stacks, VMware releases a new version of the impacted buildpacks and stacks to provide the latest patched version of the software that they reference.

The most recommended way to stay up to date with TBS component versions is to enable the automatic update ability that TBS is equipped with. When we deploy TBS with our Tanzu Network account credentials, TBS deploys a CRD object in our Kubernetes cluster named **TanzuNetDependencyUpdater**. This CRD object is responsible for keeping our TBS components up to date automatically. We used the same approach in our installation, which we performed earlier in this chapter. You can verify this setup by running the following command:

```
$ kubectl get TanzuNetDependencyUpdater -A
  NAMESPACE        NAME                 DESCRIPTORVERSION    READY
  build-service    dependency-updater   100.0.240            True
```

Here, the value of DESCRIPTIONVERSION may be different, depending on the latest available release of the description file that you would have downloaded as a part of the prerequisites at the beginning of this chapter.

If the automatic update was not enabled during the installation process, then the following link shows how to enable it post-installation or how to manually update various components to retain more control over time and the impact of the changes: https://docs.vmware.com/en/Tanzu-Build-Service/1.3/vmware-tanzu-build-service-v13/GUID-updating-deps.html.

Managing images and builds

There are various day-to-day operations that we may need to perform to work with the application configuration, in the form of image objects, and their corresponding build processes, created in the form of build objects, that are triggered for different possible reasons we saw earlier, including STACK, BUILDPACK, CONFIG, COMMIT, and TRIGGER. You can learn more about such operations at https://docs.vmware.com/en/Tanzu-Build-Service/1.3/vmware-tanzu-build-service-v13/GUID-managing-images.html. Additionally, TBS keeps the last 10 successful and failed pieces of build history information, including their completed pods and hence the logs for each Image resource. Such historical builds help you obtain historical logs and details but also occupy a lot of storage space on the cluster. In a large enterprise-scale environment, it could impact more because of several `Image` resources being created. You may refer to this documentation if you want to change the default configuration of 10 historical builds to a different number: `https://docs.vmware.com/en/Tanzu-Build-Service/1.3/vmware-tanzu-build-service-v13/GUID-faq.html#faq-18`.

Configuring role-based access controls

It is recommended to install TBS on a Kubernetes cluster that is dedicated to such platform services that are different from the cluster running actual business workloads. Such supporting services clusters are under the control of a specific user group of platform operators. Such cluster-level separation is one good way to selectively allow users to access TBS in the first place. An accidental change in a `ClusterBuilder` definition may cause a trigger to build possibly hundreds of container images for the applications that are linked with that `ClusterBuilder`. And if there is an automated deployment pipeline in place that deploys new versions of all the applications with the new container images, then such a mistake could be even more severe. That is why putting the required guardrail around TBS is very important. For that reason, TBS provides some level of access control using two Kubernetes ClusterRoles, as follows:

- `build-service-user-role`: To allow working with images and build resources
- `build-service-admin-role`: To allow all other administrative activities on TBS

The TBS users with access to images and builds should create these objects in their respective Kubernetes namespace to restrict access to these objects to the members of the same namespace. This way, we can combine the power of Kubernetes access control capabilities for greater control. You may find more details on how to configure these permissions here: `https://docs.vmware.com/en/Tanzu-Build-Service/1.3/vmware-tanzu-build-service-v13/GUID-tbs-with-projects.html#rbac-support-in-tanzu-build-service`.

Upgrading TBS to a newer version

Upgrading TBS to a newer version is very simple. You just need to perform the same steps that we walked through for the installation process other than re-importing the dependencies if they're not required.

Uninstalling TBS

To uninstall TBS from your Kubernetes cluster, just run the following `kapp` command; it will delete all TBS objects from your cluster other than the container images created by TBS. This command is very destructive and should be used with extreme caution:

```
$ kapp delete -a tanzu-build-service
```

Customizing buildpacks

TBS is built with a very modular, customizable, and extendable architecture. It allows us to perform various custom changes, such as including new buildpacks, changing buildpack order, adding new OS layers, and many more. You may learn more about such customizations using the following references:

- Managing `ClusterStores`: `https://docs.vmware.com/en/Tanzu-Build-Service/1.3/vmware-tanzu-build-service-v13/GUID-managing-stores.html`

- Managing `ClusterStacks`: `https://docs.vmware.com/en/Tanzu-Build-Service/1.3/vmware-tanzu-build-service-v13/GUID-managing-stacks.html`

- Creating and managing buildpacks: `https://buildpacks.io/docs/buildpack-author-guide/`

Summary

In this chapter, we learned about different problems around building secured container images for our applications and how TBS targets them with different capabilities. Later, we took a deep dive into the full anatomy of TBS to understand all its building blocks. After that, we walked through the installation process of TBS to get started with it. And finally, we saw how to perform various key operations on TBS. Using a solution such as TBS that is based on CNB is one of the most recommended approaches suggested by CNCF, and now, we can appreciate why that is the case.

In any Kubernetes environment, we deploy two different types of container images – either they belong to our application or a third-party software. Now that we've learned how to build secured container images for our applications using an out-of-box automation tool, in the next chapter, we will learn how to consume secured container images of popular open source software to provide backing services to our applications.

4

Provisioning Backing Services for Applications

In the previous chapter, we saw how to build application container images with ample security and without much operational overhead using **Tanzu Build Service** (**TBS**). These container images are the essential building blocks to run our cloud-native applications on container orchestration platforms such as **Kubernetes**. We can deploy those container images on a Kubernetes cluster and run our applications. However, in real life, things are not that straightforward. In the majority of cases, business applications depend on backing services such as databases, queues, caches, and others. Additionally, there is an increasing trend to also deploy such off-the-shelf backing services as containers on Kubernetes-like platforms for various good reasons.

In this chapter, we will take a deep dive into **VMware Application Catalog** (**VAC**), which provides a secure, fast, and reliable way to use such open source backing services in a containerized environment. We will cover the following topics:

- Why VMware Application Catalog?
- What VMware Application Catalog is
- Getting started with VMware Application Catalog
- Common day-two activities with VMware Application Catalog

We have a lot of ground to cover. So, let's get started exploring what business and technical challenges are addressed by VAC.

Technical requirements

There are some technical requirements that need to be fulfilled before we start using VAC. These requirements are covered later in this chapter, at the beginning of the *Getting started with VMware Application Catalog* section. However, you may not need them to understand the application and the details of this tool.

Why VMware Application Catalog?

The following are the key areas where VAC addresses detailed challenges with its capabilities for delivering better developer productivity, security, and operational practices when it comes to providing a way to consume popular **open source software (OSS)** and deploy it as running containers.

Using the right tool for the right purpose with the flexibility of choice

As we discussed previously, most business applications depend on one or more backing services, and depending on the nature of the application, the need for such backing services can be different. We have seen that using a relational database as a backend data store has been the most common backing service for the past several years. But some modern cloud-native applications could perform better with other data stores such as NoSQL. Similarly, if an application needs a queue as a backing service, we can use either Kafka or RabbitMQ. But both Kafka and RabbitMQ have their own niche use cases where one might be a better option than the other depending on the application's needs. Similarly, such options exist for tools such as caches, logging, **continuous integration/continuous deployment (CI/CD)** automation, and many other aspects of running cloud-native applications. For these use cases, there are strong and mature open source software solutions available today that are very popular and widely adopted. *Figure 4.1* shows how OSS tools have become more popular than proprietary tools in the recent past:

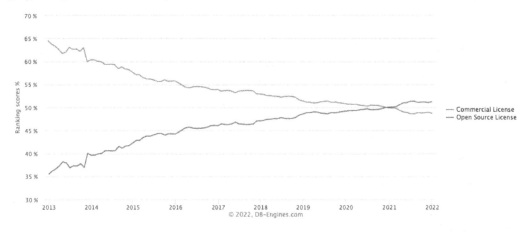

Figure 4.1 – The increasing trend of open source data store popularity

Despite the vast choice and proven track record, providing the required freedom to adopt such OSS tools internally to application teams is often challenging. There are two possible reasons for this. Firstly, it takes a lot of operational overhead to get such tools approved for usage and create a secure supply chain of container images for these tools. And secondly, the fact that their respective container images are available on public container repositories means that they are not always trustworthy. Because of that, development teams suffer either because of a lack of choice or the loss of productivity results in wait times.

VAC addresses this challenge by providing a huge catalog of OSS tools that enterprises can select from. Once a custom catalog of OSS tools is prepared, the VAC service creates an automated supply chain to stream container images and Helm charts of those selected tools and delivers them to a targeted container registry that is deployed either internally or externally, such as **Google Container Registry (GCR)**. Later, we would keep getting the newer versions of those OSS tools in the form of their newer container images and Helm charts. These can all be configured with minimal operational overhead and provide a lot of choices for application teams. Once we start getting the artifacts, container images, and Helm charts for our catalog items, we can expose that catalog for internal consumption. Then, the authorized internal members may use the catalog to provision those OSS tools on Kubernetes within minutes. Such flexibility of choice without the operational overhead to developers should encourage the usage of the right backing services for the right use case without affecting the productivity of their users.

Enhanced security and transparency

Using OSS deployment on container platforms is very quick, easy, and becoming popular. All major public container repositories such as Docker Hub and GCR have the container images for all major OSS. We can simply provide the name of the container image and download it in seconds. However, despite several benefits of pulling container images from such public container repositories, almost all enterprises, with some level of mature security practices, would not allow this. The following are a few reasons for this:

- It is difficult to determine which container images on public repositories are hosted by legitimate sources. You may find several different container images for the same software hosted by different organizations or common users. In this case, it would be very difficult to create a whitelist of authenticated sources to pull the images hosted by them.

- Even if there is a way to find the authenticated sources whose container images can be used, it is very difficult to create a governance model around it to restrict the usage of the container images hosted by unknown sources.

- It is also very difficult to get a **bill of material** (**BOM**) to know what such externally provided container images contain. Such third-party images often act as black boxes for enterprises to obtain the required confidence to allow using them in any environment. Auditing the environments running such black-box container images would be very difficult.

- Any reasonably mature organization has a standard set of **operating systems** (**OSes**) that they allow in their infrastructure. These OS requirements are often applicable to container images too. However, there are no controls and enough choices for selecting the OS when it comes to the container images provided by third parties. This could be a single significant reason that corporates disapprove of the use of the container images of OSS tools hosted on public container registries.

- There are several audit and compliance requirements when it comes to information security standards for a security-first mindset organization. To achieve the required confidence to run an OSS container in the production environment of a corporate entity, there should be several details available, including **Common Vulnerabilities and Exposures** (**CVE**) scan reports, test results, anti-virus scans, and other details for all the workloads deployed.

To meet these requirements, companies usually take control of curating their own OSS container image-building process. And when they add more OSS into their catalog of the in-house image curation process, they often later realize that such efforts are not very scalable, fast enough to fulfill the demand, efficient, or secure. Here are a few reasons for this: when there are new OSS tools required to be added to the internally managed catalog, it requires building a new automation pipeline, test cases, infrastructure capacity, and more:

- Such in-house and bespoke automation efforts are often understood and maintained by very few engineers. Out rotation of such key people in the team creates a vacuum that is difficult to fill at times, and that creates a knowledge gap.

- As the catalog becomes bigger, the maintenance effort increases exponentially. Because every newer version of each catalog item requires building a new corresponding container image for internal consumption. Because of such added overhead, the platform operations team may fall behind in keeping up with the latest patched versions of the OSS tools. Such delays in producing the latest patched container image of the tool increase security risks by allowing unpatched CVEs to be available for enterprise-wide consumption until the newer container image is ready.

- Because of the amount of effort required to provision a new catalog item for internal consumption, there could be a potential pushback to adding new items. Such pushbacks could either reduce developer productivity as they spend time waiting or affect the business application's quality by not using the right tool for the right use case.

- Even though the platform team agrees to add a new item to the catalog, it would take a long time before the actual consumers would get a production-certified container image that they can use. Such delays again waste the valuable productive time of an important workforce. Or the workforce finds workarounds by using externally available but insecure sources for such container images.

- The time of the people working on such internal automation efforts and the infrastructure capacity utilized for this reason could be better used for a more business outcome-driven endeavor.

To address these challenges, VAC comes into the picture and helps with the following benefits to provide a secure solution to increase developer productivity and operational efficiency:

- VAC allows enterprises to use their own golden OS image layer for all their selected OSS container images. This is a significant benefit as the client organization can use their hardened OS layer with desired security configurations on their selection of OS flavor.

- VAC creates an automation pipeline for the creation and distribution of every catalog item's container image and Helm chart (if applicable). Because of such automation, VAC can quickly supply newer patched versions of the catalog items to subscribers soon after the newer upstream version becomes available. Such a quick supply of the patched versions provides a good preventative security posture against hacking attacks.

- VAC supplies the following artifacts with all container images delivered to enhance consumers' confidence and increase the transparency of the container images:

  - Asset specification detailing information about the content of the container image

  - Automation test case results for a test run executed before delivery

  - CVE scan report

  - CVE scan report in CVRF format

  - Antivirus scan report

  - User documentation references

- VAC pushes all the artifacts to a private container registry as provided by the clients, which creates a trustable source of all container images and Helm charts for internal consumption.

What is the CVRF?

The **Common Vulnerability Reporting Framework** (**CVRF**) Version 1.1 was released in May 2012. CVRF is an XML-based language that enables different stakeholders across different organizations to share critical security-related information in a single format, speeding up information exchange and digestion. Reference: `https://www.icasi.org/the-common-vulnerability-reporting-framework-cvrf-v1-1/`.

In the previous points, we discussed most of the important benefits of VAC. Presently, VAC is only applicable to OSS tools. However, many large organizations do not use OSS tools without enterprise-grade support, especially in a production environment. It is important to understand that VAC, as a solution, only supports the secure supply chain of OSS container images and Helm charts using client-selected OS layers. But VAC does not support the underlying OSS tools as a part of the subscription. For example, if an organization requires PostgreSQL DB container images via VAC, then VAC will support the packaging and the timely distribution of the container images of PostgreSQL DB upstream versions. But VAC would not support the underlying PostgreSQL DB itself. Hence, for the support of PostgreSQL, the enterprise may need to either use a vendor-supported offering such as VMware's **Tanzu** data management service subscription, which supports open source PostgreSQL and MySQL DB. Alternatively, they could use a vendor-specific flavor of the open source solution such as the one provided by Crunchy Data for PostgreSQL, for example. In this case, organizations may get container images from respective third-party vendors such as Crunchy Data. For such cases, VAC would not be useful. But if the enterprises wanted to use the vanilla upstream version of PostgreSQL, which is supported by a vendor such as VMware or **EnterpriseDB** (**EDB**), then they may use VAC to benefit from all the listed benefits.

> **Upstream versus a vendor-specific flavored OSS**
>
> Using upstream OSS distributions directly without adding a vendor-specific flavor helps avoid potential vendor lock-ins, which is the first and foremost reason to use OSS technology. VAC makes the adoption of such OSS tools easier for enterprises. Despite that, many organizations still use vendor-specific flavors of OSS because of the additional features and functionalities not available in the upstream OSS distributions. Hence, there are pros and cons to both approaches.

After understanding how and where VAC is beneficial and where it is not applicable, let's now take a deeper look into what VAC is and what it contains.

What VMware Application Catalog is

After delivering comprehensive detail on what business, security, and technological challenges VAC can address and where it will not be a good use case, let's now understand this tool in a bit more detail to see what it contains. But before that, let's look into the background of VAC to learn more about it.

The history of VMware Application Catalog

In late 2019, with a vision of curating a comprehensive portfolio of modern application development and management tools, VMware decided to acquire a popular OSS packaging and distribution company named **Bitnami**. Bitnami had several years of experience working in this space, initially providing well-curated and consumable OSS tools in the form of binaries, virtual machine images, and container images. After the acquisition, VMware rebranded Bitnami as **Tanzu Application Catalog** to define an enterprise-grade OSS container image distribution offering that can customize the image specification as per the enterprise client's needs. In 2021, VMware decided to also include the **Open Virtual Appliance** (**OVA**) image catalog to build virtual machines in addition to just the container images and Helm charts, which was the original idea behind Tanzu Application Catalog. As we see in this book, VMware's Tanzu product portfolio contains all the tools and technologies around containers and Kubernetes. This was the reason why this offering was also initially given a *Tanzu* name. But with the recent announcement by VMware to also expand this offering to cover OVA images along with container images and Helm charts, this offering was renamed **VMware Application Catalog** (**VAC**), as it is not just about the container ecosystem anymore.

> **Important Note**
>
> The main focus of this book is Tanzu and its surrounding ecosystem. Additionally, at the time of writing this book, the offerings around virtual machine images are still evolving to get to the level of container images with respect to VAC. Hence, we will only cover details around container image catalog management and consumption in this chapter and not for virtual machine images.

After tapping into the history of VAC and knowing why it has *VMware* and not *Tanzu* in the name, let's now understand what the key parts of this product offering are.

Components of VMware Application Catalog

The following tools are the components of VAC. Let's review them.

VMware Application Catalog portal

The main component of this offering is the VAC portal, where we can curate and manage the application catalog. It's a **Software as a Service (SaaS)** component that is hosted and managed by VMware. VAC clients can use this portal using their VMware Cloud (`https://clould.vmware.com/`) account to access the VAC service. A catalog administrator may create new catalogs, add new OSS offerings in the catalog, and download supporting elements related to the catalog items, including test result logs, CVE reports, anti-virus scan reports, and other such items. In summary, the VAC portal provides a web-based user interface for securely curating a catalog of OSS tools that can be freely used by the internal users of the enterprise. We will cover more details about this portal later in the chapter.

Kubeapps

Once a catalog administrator defines a catalog of supported OSS tools with Helm charts on the VAC portal, developers or operators may use Kubeapps (`https://kubeapps.com/`) to consume the published catalog items for internal use. Kubeapps is an OSS tool under the **Cloud Native Computing Foundation (CNCF)** umbrella. This project was started by Bitnami to provide a **graphical user interface (GUI)** to deploy software using Helm charts on top of Kubernetes clusters. Since the acquisition of Bitnami, VMware actively maintains it. It is a very lightweight application that can be deployed on a Kubernetes cluster running in the organization's environment. The users of Kubeapps can select software to be deployed from the accessible catalog, change required deployment configurations (for example, user credentials, storage size, security configuration, and things of that nature), and finally deploy it as running containers on the targeted namespace of the selected Kubernetes cluster. Once a new version of the software is available in the catalog, the user can quickly upgrade it or remove the deployment if no longer required. To sum up, if the VAC portal provides the required controls for securely exposing a catalog of OSS tools, the Kubeapps provides the desired flexibility and productivity to developers or other users of the catalogs to life cycle various OSS tools as and when required using a published catalog.

Let's understand how these components work together to provide the required functionalities. *Figure 4.2* describes the overall process to define the OSS catalog on the VAC portal and consume those catalog artifacts either using Kubeapps or **continuous deployment (CD)** automation in its place.

The VAC management and consumption process

After reviewing the key components of this offering in the previous section, let's now understand how everything works together to provide end-to-end functionality of catalog curation and consumption using VAC. The following points correspond to each number given in *Figure 4.2* and describe what happens during that step of the process. The cloud in *Figure 4.2* depicts the SaaS infrastructure of VAC, and the rectangle defines the infrastructure boundary of the client organization. This client infrastructure may be either a private data center, a public cloud, or a combination of both:

1. A catalog administrator within the client organization with access to the VMware Cloud account defines a new catalog of selected OSS tools. In this step, the catalog admin also selects the base container OS to be used, whether to include or exclude Helm charts for the selected items and provides a container repository reference to where the catalog artifacts will be pushed for secure consumption.

2. Once the catalog administrator submits a catalog configuration, VAC automation processes take it forward to deploy the required automation pipelines to create a stream of container images and Helm charts that can be generated as per the specifications provided by the catalog administrator in *step 1*.

3. The VAC automation process pulls required OSS binaries from authorized third-party sources based on each selected OSS tool. After getting the required binaries, VAC automation performs certain operations, including automation testing of the OSS tool's version, packaging the tool using the client-specified OS image layer, preparing a container specification report, and performing anti-virus and CVE scans. This step gets repeated for each newer version of each OSS tool covered by VAC.

4. Once a catalog item is ready for consumption, it is pushed to a target container registry as specified by the catalog administrator in *step 1*. This step is also repeated for each OSS tool's version that is prepared in *step 3*.

5. Once a catalog item is pushed in the target container registry, a catalog administrator may pull the required reports that were prepared in *step 3* for the published artifacts such as CVE scan reports and others. The VAC portal also specifies required CLI commands to use the container images and Helm charts for the published artifacts.

6. In this step, the catalog administrator configures the published catalog on Kubeapps for internal consumption.

7. In this step, Kubeapps pulls the required details of the catalog to publish on the GUI for consumption.

8. Once the catalog is configured on Kubeapps, a catalog consumer may access Kubeapps GUI to deploy the required OSS tool as a Kubernetes deployment:

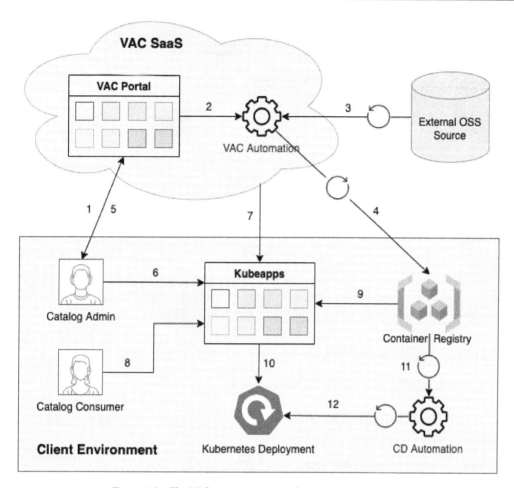

Figure 4.2 – The VAC management and consumption process

9. Upon receiving the request from a catalog consumer to deploy a tool using its Helm chart on the Kubeapps GUI, Kubeapps pulls the required Helm chart and container images for the deployment of the tool from the container registry where the artifacts were pushed in *step 4*.

10. The Helm installer triggered by Kubeapps deploys the OSS tool on the targeted Kubernetes cluster using the configuration supplied by the catalog consumer in *step 8*. At the end of this step, we have a running instance of the OSS tool in the targeted Kubernetes cluster. In most cases, this step gets completed within a few minutes.

11. This step describes an alternative way of consuming the VAC-supplied artifacts using a CD automation process. This step can be configured to be triggered every time there is a newer version of the artifact available in the container registry to initiate an automated deployment process.

12. In this step, the CD process deploys the downloaded OSS tool in the targeted Kubernetes cluster.

The following figure shows all these steps as a summary highlighting what VAC is and how it is used:

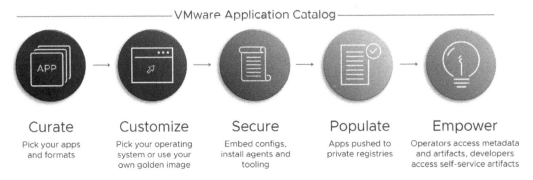

Figure 4.3 – VAC at a high level (https://docs.vmware.com/)

With this, we covered the details required to understand the history of VAC, how VAC got its current name, the key components of VAC, and the end-to-end process involving catalog creation to consumption. Now let's get started using VAC to better understand how to consume it.

Getting started with VMware Application Catalog

In this part of the chapter, we will cover the following details:

- How to configure an application catalog
- How to install Kubeapps on a Kubernetes cluster
- How to configure Kubeapps to use a catalog

So, let's get started with hands-on work. But before that, we need the following prerequisites fulfilled.

Prerequisites

The following points list prerequisites to operationalize VAC:

- A VMware Cloud Services (`https://console.cloud.vmware.com/`) account with VAC access
- One of the following container repositories that can be accessed by VAC to push catalog items:
  - GCR
  - Azure Container Registry
  - Harbor

- A Kubernetes cluster with the following attributes:

  - Version 1.19 or later

  - Outbound internet access

  - Container registry access that is used by VAC

  - Automated **Persistent Volume (PV)** creation based on **Persistent Volume Claims (PVC)**

- A workstation with either Linux or macOS

- Helm v3.x installed on the workstation machine

> **Helm installation**
>
> Use this documentation if you need help with the Helm installation: `https://docs.bitnami.com/kubernetes/get-started-kubernetes/#step-4-install-and-configure-helm`.

Let's start by defining a catalog of backend services that can be accessed by various applications. We will select MySQL, a relational database, as an example of an OSS catalog item to describe various details later in the chapter. In real life, we may add many other OSS tools to the catalog and use them in a similar way. You can find a broader list of available OSS tools on the VAC portal for catalog creation.

Creating a catalog on the VAC portal

To create a catalog on the VAC portal, take the following steps:

1. Log in to your VMware Cloud Services account and select **VMware Application Catalog** from the available services. If you do not see that service listed, then you may need to reach out to your VMware account team member to get access:

Figure 4.4 – Select VAC from the list of services

2. You will see an empty catalog page, as shown in the following figure as there will be no catalog items previously added. Click on the **ADD NEW APPLICATIONS** button shown in the following figure:

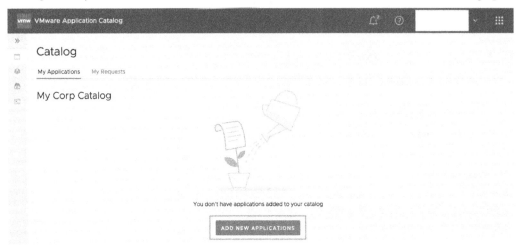

Figure 4.5 – The ADD NEW APPLICATIONS button in an empty catalog

3. Select the base OS layer for the catalog artifacts. As mentioned before in this chapter, we will focus on the Kubernetes-based application catalog and not on virtual machines. As a simple example, **Ubuntu 18.04** is selected as the base OS layer, but you may also select the **Custom Base Image** option. For more details on that visit the product documentation for VAC at `https://docs. vmware.com/en/VMware-Application-Catalog/services/main/GUID- get-started-get-started-vmware-application-catalog.html#step-3- create-custom-catalogs-5`:

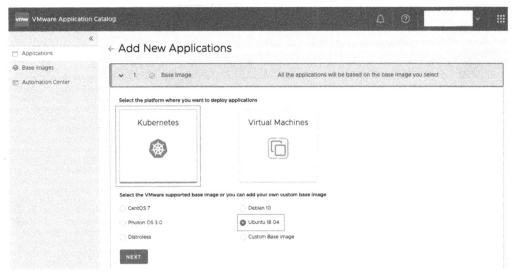

Figure 4.6 – Base OS selection for catalog items

4. Select the required OSS items from the available options to include in the catalog. You may also search for them if required:

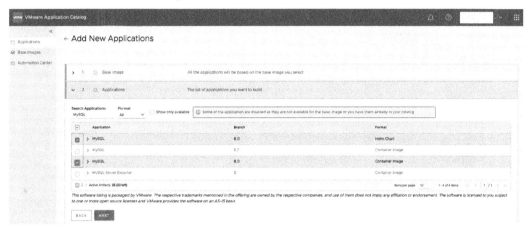

Figure 4.7 – Select catalog items

5. Add the destination container registry by clicking on the **ADD REGISTRY** button to get the required catalog artifacts delivered:

Figure 4.8 – Add the container registry

6. Provide the required details for the container registry to allow VAC to push catalog artifacts to it:

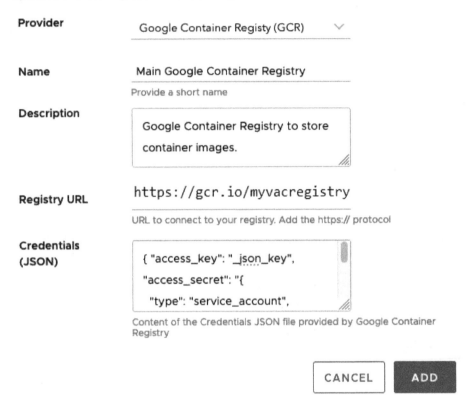

Add Registry

Registries are used to push newly requested applications, application updates and storing custom base images

Provider   Google Container Registy (GCR)   ⌄

Name   Main Google Container Registry

Provide a short name

Description   Google Container Registry to store container images.

Registry URL   https://gcr.io/myvacregistry

URL to connect to your registry. Add the https:// protocol

Credentials (JSON)   { "access_key": "_json_key",

"access_secret": "{

"type": "service_account",

Content of the Credentials JSON file provided by Google Container Registry

CANCEL   ADD

Figure 4.9 – Add registry details

Though *Figure 4.9* shows the details for GCR, it also supports other registries, including Azure Container Registry and Harbor. You can get more information regarding other registries at: `https://docs.vmware.com/en/VMware-Application-Catalog/services/main/GUID-get-started-get-started-vmware-application-catalog.html#step-3-create-custom-catalogs-5`.

7. Give a name and description to this catalog:

Figure 4.10 – Add name and description for the catalog

8. Verify the input summary and submit the catalog request:

Applications	Base Image	Registry
MySQL (Chart)	Ubuntu 18.04	VMware will reach out to gather more details about your registry
MySQL (Container)		

Figure 4.11 – Submit the catalog request

After this, we will see a message confirming our catalog request submission. It may take about a couple of weeks to review and process this request before we start getting our catalog items delivered to our selected container registry destination.

That's it. We have our catalog of OSS tools defined so that we can publish to internal consumers for easy and quick access. But how can they access this catalog? Let's check that out in the next section.

Consuming VAC using Kubeapps

In the previous section, we learned how to request an application catalog on the VAC portal. Once we start getting our Helm charts and container images delivered in our container registry, we can access those tools using Kubeapps, a GUI to manage Kubernetes-based software deployments. We discussed Kubeapps in detail in the *Components of VAC* section covered previously in this chapter. Let's see how to install and configure it now.

Kubeapps installation

Kubeapps is itself a Kubernetes-based deployment to manage other Kubernetes deployments that can be deployed using Helm charts and operators. Since we are planning to use Kubeapps to consume our VAC-supplied distributions that include Helm charts and container images, we will not cover the usage of operators in this section. As a deployment topology, we can install Kubeapps on any Kubernetes cluster to deploy catalog items on the same or any other Kubernetes clusters and their Kubernetes namespaces that are linked with Kubeapps. In this chapter, we will use a single Kubernetes cluster to minimize configuration complexity.

We will need to fulfill some requirements to move forward, as they are covered previously in the *Prerequisites* section of this chapter. The following steps describe the installation and configuration of Kubeapps on a Kubernetes cluster with sufficient resources:

1. Add a Bitnami Helm chart repository to your local Helm library:

    ```
    $ helm repo add bitnami https://charts.bitnami.com/
    bitnami
    ```

2. Create a Kubernetes namespace for Kubeapps:

    ```
    $ kubectl create namespace kubeapps
    namespace/kubeapps created
    ```

3. Install Kubeapps in the kubeapps namespace using a Helm chart:

    ```
    $ helm install kubeapps --namespace kubeapps bitnami/
    kubeapps
    NAME: kubeapps
    LAST DEPLOYED: Wed Jan 19 19:45:20 2022
    NAMESPACE: kubeapps
    STATUS: deployed
    REVISION: 1
    TEST SUITE: None
    NOTES:
    CHART NAME: kubeapps
    CHART VERSION: 7.7.0
    . . .

    . . .
    ```

    The preceding output is truncated for conciseness. With this, we have Kubeapps running in our cluster. Let's verify it.

4. Verify the Kubeapps deployment to see if everything is running fine:

    ```
    $ kubectl get all -n kubeapps
    ```

    The preceding command should list pods, services, deployments, replicasets, statefulsets, and jobs that are not listed here for brevity.

5. Create a temporary service account to access the Kubeapps GUI:

    ```
    $ kubectl create --namespace default serviceaccount
    kubeapps-operator
    serviceaccount/kubeapps-operator created
    ```

6. Link the service account with a Kubernetes role to allow required access:

    ```
    $ kubectl create clusterrolebinding kubeapps-
    operator --clusterrole=cluster-admin
    --serviceaccount=default:kubeapps-operator
    clusterrolebinding.rbac.authorization.k8s.io/kubeapps-
    operator created
    ```

7. Retrieve the access token for the account to log in to the Kubeapps GUI:

    ```
    $ kubectl get --namespace default secret $(kubectl get
    --namespace default serviceaccount kubeapps-operator -o
    jsonpath='{range .secrets[*]}{.name}{"\n"}{end}' | grep
    kubeapps-operator-token) -o jsonpath='{.data.token}' -o
    go-template='{{.data.token | base64decode}}' && echo
    ```

    This long command will print a long string of characters, which is the token that we will use to log in to the Kubeapps GUI. Save this token for future reference.

Important note

The previous three commands to create access permissions for Kubeapps use a very primitive approach for simplicity and as an easy learning reference. In a production-grade implementation, we may need to use a more sophisticated approach to configure real enterprise users who can access Kubeapps in the Kubernetes namespace. Such user permissions are generally managed using an external integration with either an OIDC or LDAP identity provider. Additionally, using the cluster-admin role is not a secure approach and should not be used other than for such learning purposes.

8. Expose the Kubeapps GUI to access it locally in a browser. For a production-grade deployment, the GUI should be assigned a proper domain name and exposed outside the Kubernetes cluster using a load balancer. In this step, we will use Kubernetes port forwarding for quick and simple access to the deployment:

```
$ kubectl port-forward -n kubeapps svc/kubeapps 8080:80
Forwarding from 127.0.0.1:8080 -> 8080
Forwarding from [::1]:8080 -> 8080
```

9. Access Kubeapps in your local browser using `http://localhost:8080/`. This should open the following screen of Kubeapps:

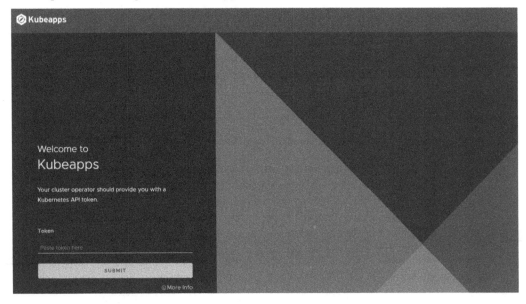

Figure 4.12 – Kubeapps GUI – authentication page

10. Access Kubeapps using the token retrieved in *step 7*. Paste the token value on the login page and submit. This should open Kubeapps GUI as shown in *Figure 4.13*:

Figure 4.13 – Kubeapps GUI – initial landing page

> **Important note**
>
> The screenshots and steps described in this chapter are based on the presently available versions of VAC and Kubeapps. Depending on when this book is read, the content and experience could be different based on the future changes incorporated into these products.

As you will see, Kubeapps comes with a configuration to access the publicly available generic Bitnami catalog. Once our customer catalog defined on the VAC portal is ready, and we start getting its artifacts, we may configure Kubeapps to use the same. We will cover the linking of our custom application catalog on Kubeapps later in day-two activities.

With this, we conclude this section of the chapter on getting started with VAC and Kubeapps. We have seen how to create a catalog on the VAC portal and install Kubeapps in a Kubernetes cluster. In the next section, we will see the following items, covering common day-two activities around VAC:

- How to inspect delivered catalog artifacts and obtain required reports

- How to link VAC with Kubeapps to publish it for consumption

- How to consume catalog items via automation pipeline using Kubeapps

- How to deploy MySQL as a backend service running on the Kubernetes cluster using Kubeapps

- How to manage catalog items

Let's now learn how to use VAC.

Common day-two activities with VAC

With the details covered in the previous section of the chapter, we now have an OSS tool catalog request placed on the VAC portal and a Kubeapps instance running in the Kubernetes cluster. Let's now review some of the key day-two activities that can be performed by the catalog administrator to ensure security and compliance of the OSS tool usage, and by the catalog consumers to unleash productivity and flexibility to quickly deploy and use these OSS tools in different ways that are part of the VAC catalog.

Inspecting catalog deliverables

Once our catalog request is submitted to Vmware using the VAC portal, as we covered in the previous section, we can check the status of our catalog deliverables using the VAC portal as shown in the following figure under the **My Requests** tab:

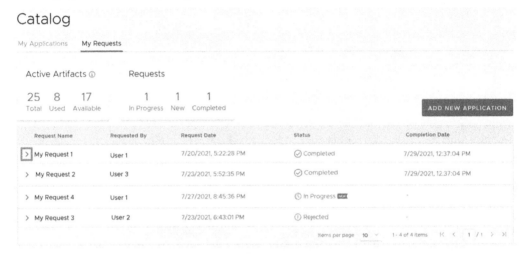

Figure 4.14 – Checking VAC request status

When we have our catalog request completed, we start seeing our delivered artifacts under the **My Applications** page, as shown in the following screenshot:

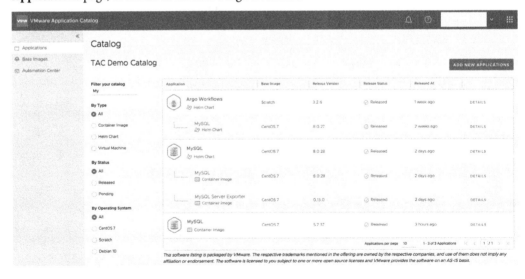

Figure 4.15 – Listing catalog applications on the VAC portal

Depending on our selection during the catalog creation request, we may see Helm charts and their container images, as shown in *Figure 4.15*. Let's now check the details of the MySQL container image by clicking on the **DETAILS** link given for the item on the right. The following screenshot shows the details for the MySQL container image:

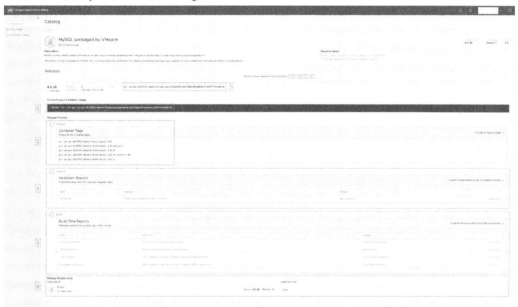

Figure 4.16 – Details of a catalog item on the VAC portal

We can get the following details from different sections of a catalog item's details page, as shown in *Figure 4.16*. The numbers in the following list correspond to the numbers given in *Figure 4.16*:

1. **Digest**: This is the name and the location of the artifact that is placed in our destination container registry that we supplied during catalog creation.

2. **Consume your Container Image**: This is the Docker command to pull this container image into a Docker runtime environment.

3. **Container Tags**: These are different tag alias for this container image that we can use in a Kubernetes deployment manifest file to pull this container image for running this application on Kubernetes.

4. **Validation Reports**: This allows us to download the automation test result log file that was generated to test this version of MySQL before creating this container image.

5. **Build Time Reports**: This section contains various container build reports, including anti-virus and CVE scan reports and an asset specification (bill of material) report containing the list of software with their versions used in the container image.

6. **Release Relationship**: This section shows the dependent Helm charts that use this image.

Like the details of a container image, the details of the corresponding Helm chart include required `helm` CLI commands to deploy the chart, test results for the chart, asset specification, and container image dependencies.

Using the application catalog

After inspecting the required details for the artifacts supplied for our catalog items on the VAC portal, it's time to use them to deploy those tools in our environment. There are various ways we can use those container images provided by VAC. Depending on the requirement of the tool, we can simply run some tools in our workstation's container runtime environment, such as Docker. However, for an enterprise-grade deployment, we need several supporting components to run a tool. For example, we saw that all Kubernetes resources were created to deploy and run our Kubeapps instance that we deployed previously in this chapter. One possible way to deploy such tools is to use deployment automation using custom scripts and CI/CD tools such as Jenkins and/or **ArgoCD** using a **GitOps**-based deployment model. Another possible option is to use a Kubernetes packaging tool such as Helm charts. A Helm chart bundles all required dependencies to deploy and run corresponding objects for a tool on a Kubernetes cluster. Using a tool such as Helm charts makes it very easy to configure and quickly deploy the tool with all its required components with minimal effort and within a few minutes. As we have seen previously, VAC allows us to select containers as well as Helm charts for our catalog items wherever applicable. As a part of application packaging, Helm charts also allow exposing certain configuration properties that we may need to change for different deployments. We can use these Helm charts provided by VAC to deploy those tools with our custom configuration requirements. For example, for a MySQL database deployment, we can change attributes such as its name, login credentials, storage volume size, and many more.

What is GitOps?

GitOps upholds the principle that Git is the one and only source of truth. GitOps requires the desired state of the system to be stored in version control so that anyone can view the history of changes. All changes to the desired state are performed through Git *commits*. Source: `https://blogs.vmware.com/management/2020/08/ops-powered-by-git-gitops-in-vrealize-automation.html`.

As we discussed previously, Kubeapps is a tool to deploy and manage our catalog of Helm charts that we have configured on VAC. So, let's check out how we can link those Helm charts provided by VAC with our Kubeapps instance that we deployed previously in this chapter.

Adding the application catalog to Kubeapps

The following steps describe how to configure a new catalog on our Kubeapps instance for the Helm charts provided by VAC:

1. Obtain the chart repository location where the Helm charts are located. You can find the same on the details page of a Helm chart item in your catalog on the VAC portal, as shown in the following screenshot:

Release Process

Figure 4.17 – Getting the Helm chart repository location

The highlighted URL portion is the location of all the Helm charts for our catalog that we defined on the VAC portal. Make a note of this URL as we will use it in one of the following steps:

2. Generate an API token for the VAC account on the VAC portal, which we will use to authenticate our Kubeapps instance to pull the Helm charts for selected catalog items. The following sub-steps describe how to generate an API token on the VAC portal:

   I. Go to the **My Account** page on the VAC portal using the drop-down menu in the top-right corner, as shown in the following screenshot:

Figure 4.18 – Go to the VAC My Account settings

II.   Click on the **API Tokens** tab:

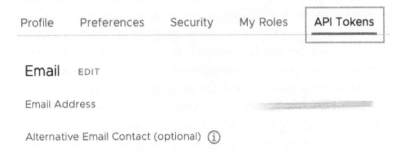

Figure 4.19 – Go to the API Tokens list page

III.   If there is already a token listed on this page that has access to the VAC service, then you can skip the following steps to generate a new API token and jump directly to *step 3* to add a repository in Kubeapps. Otherwise, click on the **GENERATE A NEW API TOKEN** link as shown in the following screenshot:

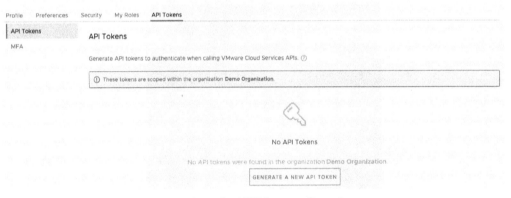

Figure 4.20 – Go to the API Tokens configuration page

IV.   Enter the token configuration, as shown in the following screenshot:

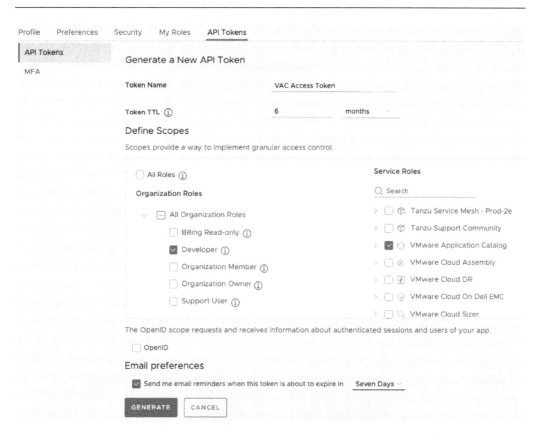

Figure 4.21 – Generate a New API Token for VAC

V.   You can change the name and duration of the token to your requirements. However, the scope is important to allow access to VAC from Kubeapps. It may not be a read-only or support role. You can also generate a generic token that can be used for all VMware Cloud Service offerings. However, it will be very broad in nature allowing all types of access to all services. Hence, it is not recommended.

VI.   Save the generated token for future usage:

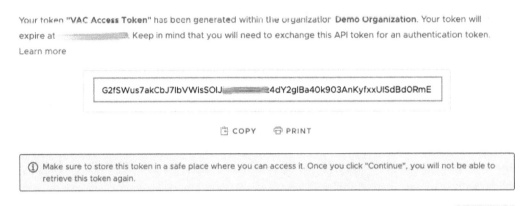

Token Generated

Your token **"VAC Access Token"** has been generated within the organization **Demo Organization**. Your token will expire at ▓▓▓▓▓▓▓▓. Keep in mind that you will need to exchange this API token for an authentication token. Learn more

G2fSWus7akCbJ7lbVWisSOIJ▓▓▓▓▓4dY2glBa40k903AnKyfxxUISdBdORmE

🗐 COPY   🖶 PRINT

ⓘ Make sure to store this token in a safe place where you can access it. Once you click "Continue", you will not be able to retrieve this token again.

CONTINUE

Figure 4.22 – Save the generated token

3. Log in to the Kubeapps instance that we deployed and open its configuration menu using the top-right corner icon followed by the **App Repositories** option as shown in the following screenshot:

Figure 4.23 – Add the Helm chart repository to Kubeapps

4. Click on the **ADD APP REPOSITORY** button to configure the details of a new repository:

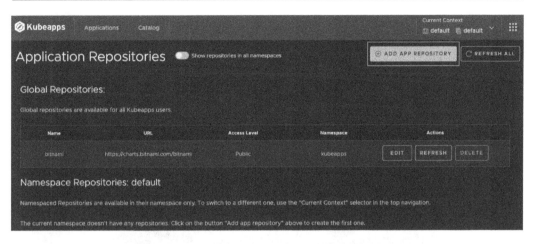

Figure 4.24 – Go to a new repository configuration page

5. Give it a name, add the Helm chart repository URL captured in *step 1*, paste the API token generated in *step 2*, select the **Skip TLS Verification** option, and click the **INSTALL REPO** button:

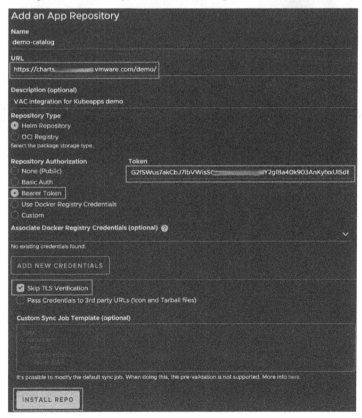

Figure 4.25 – Install the VAC Helm repo in Kubeapps

We had to select the **Skip TLS Verification** option, as shown in *Figure 4.25*, as our Kubeapps deployment is not assigned an external facing domain name and a TLS certificate. In an enterprise-grade environment, this is not a recommended approach.

6.   If you get a success message, then the catalog should be integrated with Kubeapps. To check that, click on the **Catalog** tab on Kubeapps and select the **demo-catalog** option, as shown in the following screenshot:

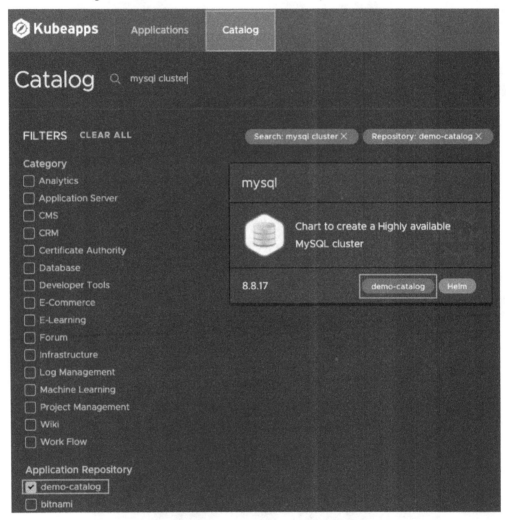

Figure 4.26 – A newly configured catalog on Kubeapps

With this, our Kubeapps deployment is fully integrated with our custom application catalog that we created on VAC. Let's now learn to deploy a MySQL service instance from our custom catalog using Kubeapps.

Deploying a service using Kubeapps

In this part, we will deploy a MySQL database instance on our Kubernetes cluster using Kubeapps. The following steps describes how to do it:

1. Click on the **MySQL** tile that is shown in *Figure 4.26* to get started with the installation on your Kubernetes cluster. We will see the following screen after clicking there:

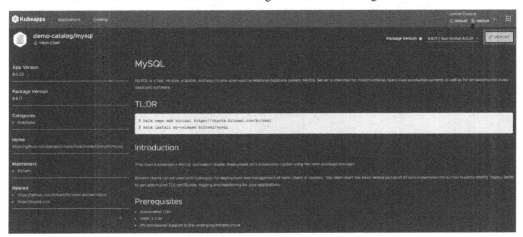

Figure 4.27 – Deployment instruction page for MySQL on Kubeapps

This page shows detailed instructions on how to use a MySQL Helm chart. We can use these instructions to manually deploy it using the Helm chart command-line tool. However, we will see how we can use the Kubeapps GUI for such configuration and installation. You can scroll down the page to see the details of all the configuration parameters that this Helm chart allows us to change to customize our MySQL database instance. By clicking the highlighted **DEPLOY** button in *Figure 4.27*, we get a screen to update these attributes for our custom needs and to trigger the installation.

2. Depending on the type of Helm chart, we may get a form like a GUI to modify some of the most common attributes for the deployment. For example, the following form in *Figure 4.28* for MySQL DB instance configuration shows that we can select deployment architecture and the size of the primary and secondary database storage volumes. But we will use a detailed YAML-based configuration approach to demonstrate that as well. So, click on the highlighted **YAML** tab in the following screenshot to move forward:

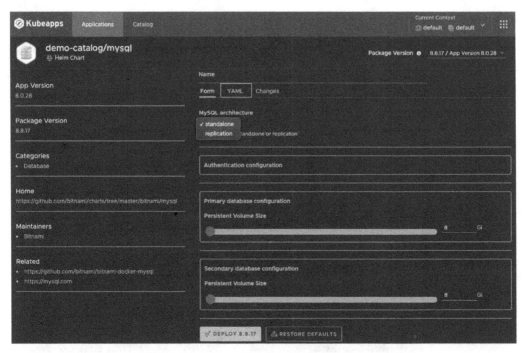

Figure 4.28 – Helm configuration form for MySQL DB on Kubeapps

3.   Select the destination Kubernetes namespace, update the **YAML** configuration, and deploy the service:

Figure 4.29 – Detailed Helm chart configuration for MySQL DB deployment using Kubeapps

- Section 1 in *Figure 4.29* shows where we can select the destination for this installation. As discussed before, we may use Kubeapps to deploy Helm charts in many other connected Kubernetes clusters and their namespaces depending on the Kubeapps configurations and the privileges of the Kubeapps user on those clusters and namespaces. In this case, we have only one option – **default** – that corresponds to the Kubernetes cluster where Kubeapps is deployed. Similarly, we may also select or create a namespace for this deployment if required.

- Section 2 in *Figure 4.29* shows the YAML configuration for the Helm chart, where we may customize our deployment configuration. The deployment used in this book has only changed for user credentials, keeping all other attributes to their default values to keep it simple. Once the required configuration changes are done, we can verify them using the **Changes** tab to ensure there are only intended changes in the deployment configuration.

- Once the required changes are made, section 3 in *Figure 4.29* shows the button to trigger the deployment on our Kubernetes cluster.

4. We will get the following page after triggering the Helm chart deployment to update the status of the installation:

Figure 4.30 – MySQL DB deployment status on Kubeapps

Figure 4.30 contains the following details:

- Section 1 shows the different buttons to perform the life cycle operations for the deployment, including upgrading it to a newer version of MySQL, rolling back to a previous version, or deleting the deployment if not required. The button triggers their corresponding Kubernetes Deployment life cycle operations behind the scenes.

- Section 2 shows the number of healthy running pods for our deployment.

- Section 3 shows the authentication credentials that were provided during the installation configuration step covered previously, as shown in *Figure 4.29*.

- Section 4 shows different useful tips and details for using the deployed tool. For example, in our case, it shows how to connect to the MySQL DB instance deployed here.

- Section 5 shows the different Kubernetes resources that were deployed as part of this installation. We can view details of the resources by clicking on their corresponding tabs.

- Section 6 shows the final installation configuration that was used to deploy this instance of MySQL DB. We can save the YAML configuration in a file for future usage to deploy the same type of MySQL instance again.

With this, we have covered the required details to see how we can use Kubeapps to consume a custom application catalog created using VAC to quickly deploy several OSS tools in minutes and in a self-service way. Such a setup given to developers to deploy required application backend services can boost their productivity, reduce overall application development time, and as a result, improve an enterprise's innovation speed. And using Kubeapps and VAC is not just for developers but also for DevOps, platform, and infrastructure engineers to deploy and use many popular OSS tools using container images and Helm charts delivered from a trusted source.

Let's now visit one more day-two activity around VAC that the catalog administrators will need to perform – updating the catalog on the VAC portal.

Updating the application catalog

Earlier in the chapter, we saw, as a part of day-two activities, how to obtain different reports and logs for the supplied artifacts from the VAC portal. Then we covered how to consume a catalog of OSS tools that were supplied by VAC using either an automation pipeline or Kubeapps. Now let's visit the last major day-two activity around VAC, which is to update the catalog items.

Adding new catalog items

To add new OSS applications to the corporate catalog, catalog administrators can go to their corporate VAC account using VMware Cloud Services credentials. Once on the VAC portal, the catalog administrators can select the **ADD NEW APPLICATIONS** button under the **Applications** section, as shown in *Figure 4.31*:

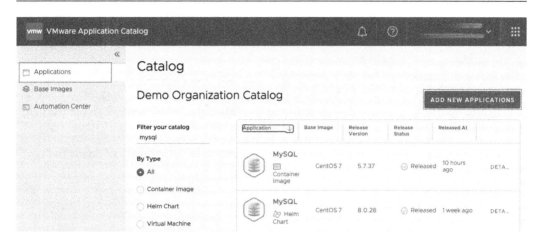

Figure 4.31 – Add a new item in the catalog on the VAC portal

The steps to add new catalog items after clicking the highlighted button in *Figure 4.31* are the same as we covered during the creation of the new catalog earlier in this chapter:

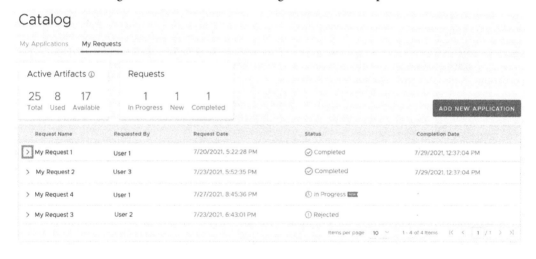

Figure 4.32 – Request status tracking on the VAC portal

Newly added items will appear under **My Requests**, as shown in *Figure 4.32*.

Making additional catalog changes

Other than adding new items in the catalog, we may also need to remove unwanted tools from it, update the base OS layer, update destination repository details, and include/exclude Helm charts from the listed items. At the time of writing this book, all these operations can only be performed by contacting VAC support as the VAC portal does not currently provide an interface to do it in a self-service manner.

With this, we have covered most of the day-two activities around VAC. Let's recap to summarize our learning from this chapter.

Summary

In this chapter, we saw that the popularity of using OSS tools has dramatically increased in recent years. There are several mature OSS tools available today that many organizations confidently use in their production environments. Also, running software such as containers on Kubernetes alongside containerized client applications is a quick way to equip the applications with appropriate backend technologies. But using container images available on public container registries such as Docker Hub is not a secure way of deploying such OSS tools. And hence, most organizations do not encourage such practices and try to employ some internal mechanisms to generate internally curated container images for such OSS consumption. Having such efforts undertaken internally not only wastes a lot of resources but also discourages developer teams from experimenting with various tools or wasting their productivity in waiting for getting them ready for consumption. And that is where VAC comes into the picture. We saw how VAC can handle such challenges by providing a way to curate a custom catalog of required OSS tools that can be consumed internally with full confidence.

We also learned how VAC works and what its key components are. Following this, we learned how we can get started with VAC to define a catalog and set up Kubeapps to consume the Helm charts and container images delivered under that custom catalog. And finally, we went through some of the key day-two activities around catalog consumption and management. We saw how quickly we can deploy an instance of MySQL database using its Helm chart supplied as a part of the catalog. We also reviewed how catalog administrators can check out and obtain copies of the CVE scanning report, anti-virus scan logs, test logs, and asset specification (bill of material) for all the container images and Helm charts delivered through the VAC service.

After seeing how to build our apps quickly with the vast choice of OSS tools to be used as backends for our apps, in the next chapter, we will learn how to build and manage the API endpoints exposed by our applications using the Tanzu application tools.

5
Defining and Managing Business APIs

Application programming interfaces (**APIs**) are as old as digital computing itself but more relevant than ever as we continue to interconnect every aspect of modern life. They make up the backbone of most enterprise applications and help run the modern world. It is this very proliferation of APIs that makes them a big source of waste, redundancy, and bad practice in enterprises.

This chapter introduces two products aimed squarely at this problem space: **Spring Cloud Gateway for Kubernetes** and **API Portal for VMware Tanzu**. As is the case with every product in the Tanzu portfolio, these products address a specific acute business need: in this case, developing, operating, publishing, securing, monitoring, documenting, searching for, and consuming APIs.

In this chapter, we will cover the following topics:

- Spring Cloud Gateway for Kubernetes and API Portal for VMware Tanzu – overview
- Why Spring Cloud Gateway for Kubernetes?
- Why API Portal for VMware Tanzu?
- Spring Cloud Gateway for Kubernetes – getting started
- API Portal for VMware Tanzu – getting started
- Spring Cloud Gateway for Kubernetes and API Portal for VMware Tanzu – real-world use case
- Spring Cloud Gateway for Kubernetes and API Portal for VMware Tanzu – day-2 operations

Spring Cloud Gateway for Kubernetes and API Portal for VMware Tanzu – overview

Before jumping into the product installation, let's revisit why we're here. I like to think of the API space in terms of three personas. There's the **API Developer**, the **API Consumer**, and the **API Operator**. Each has its own set of problems.

If I develop APIs, there are some recurring problems that I must solve with every single project:

- Exposing my API to my customers
- Terminating *TLS*
- Handling *CORS* and browser restrictions
- Making my API discoverable
- Securing my API endpoints so that only certain groups can execute certain functions
- Protecting against misuse by rate-limiting requests
- Rewriting request paths
- Rewriting request and response headers

> **API terminology**
>
> **Transport Layer Security (TLS)** is the cryptographic technology behind the reassuring lock next to your bank's website's URL in your web browser. It is especially important in the realm of APIs as most of the sensitive financial data, personal messages, and sensitive health information are delivered via an API. It's especially difficult to implement correctly, and something best handled centrally by an API gateway rather than having each API developer implement it themselves.
>
> **Cross-Origin Resource Sharing (CORS)** is an especially important topic for engineers in the API space. CORS is the mechanism by which your browser allows one website to make calls in the background (often called AJAX calls) to another website. There are legitimate reasons to do this, but this technique is often used by bad actors to try to steal credentials or sensitive information. Configuring a website and its supporting APIs such that only trusted AJAX calls are allowed is complex and easy to get wrong. This is one more reason why it's best to centralize this logic in an API gateway.

Now, let's say that I need to consume APIs. I have an entirely different set of problems:

- Does the API I need exist?
- Where can I find the APIs I need?
- How do I access those APIs once I've found them?
- Do those APIs require authentication? If so, what scopes?
- How do I test out an API before writing a bunch of code to consume it?

Finally, if I'm a platform operator hosting APIs, I may need to do the following:

- Provide developers self-service access to an API gateway that they can configure themselves
- Monitor and alert on metrics around the APIs

- Provide an API gateway with extremely high throughput so as not to become a bottleneck

- Provide a uniform implementation of common features such as the following:

  - **SSO (Single Sign-On)**

  - Rate limiting

  - Header manipulation

  - Header/content enrichment

- Where necessary, let developers build custom filters and plug them into an API gateway

- Manage the life cycle and upgrades of developers' API gateways

- Where possible, auto-generate API documentation based on what is deployed to the gateway

- Provide a single searchable location where developers can discover and try out all the APIs I manage

Spring Cloud Gateway for Kubernetes and **API Portal for VMware Tanzu** exist specifically to provide a simple, straightforward, enterprise-wide solution for all three personas' concerns.

Now that we have a high-level overview of the topics being covered, let's break it down into the "whys" of both products. We'll start with *Spring Cloud Gateway for Kubernetes*.

Why Spring Cloud Gateway for Kubernetes?

Spring Cloud Gateway for Kubernetes is based on the open source *Spring Cloud Gateway* project: `https://spring.io/projects/spring-cloud-gateway`. Spring Cloud Gateway is a library for building high-performance APIs. You deploy it like a normal Spring app and configure it like you would configure a Spring app. Unfortunately, the open source project doesn't do much to address many of the problems mentioned previously – problems commonly encountered in the enterprise.

In addition to their business logic, developers must also package, configure, and deploy a Spring Cloud Gateway app, or bundle it into their existing app as a library. Either way, it's a significant amount of added complexity. Unless they're using some advanced features of the Spring Framework, any changes to their API's routes will involve rebuilding and redeploying the app.

Furthermore, the open source Spring Cloud Gateway leaves some of the operator's problems unsolved. If every development team is doing its custom deployment of Spring Cloud Gateway, the operator can't reason about how each gateway's routes will be exposed, or how it will emit metrics. In addition, there's no easy way for multiple teams to share a gateway as they all need to configure and deploy it on their schedule.

This is where Spring Cloud Gateway for Kubernetes enters the picture. Spring Cloud Gateway for Kubernetes is a commercial Tanzu product with a superset of the open source's features. It is geared specifically toward the enterprise and managing software at scale. It uses a Kubernetes Operator to manage three main objects:

- Instances of Spring Cloud Gateway
- API routes – instructions to the gateway on how to filter incoming requests and where to send them on to
- Mappings of API routes to Spring Cloud Gateway

By exposing these three entities as Kubernetes **Custom Resources**, Spring Cloud Gateway for Kubernetes can abstract away the following:

- Packaging and deploying Spring Cloud Gateway instances
- Managing the Spring code to configure those instances
- Dynamically updating that configuration without redeploying them
- Life cycle-managing all the gateway instances across all Kubernetes clusters
- Reasoning about the gateway instances in bulk:
  - They all emit metrics the same way, so I can easily aggregate all their metrics
  - They all expose their API configuration the same way, so I can aggregate that config and report on it or, as we will see shortly, use it to drive other useful tools such as API Portal for VMware Tanzu

To summarize then, Spring Cloud Gateway for Kubernetes uses a Kubernetes Operator and some very strategic **Custom Resource Definitions** to make the already formidable open source Spring Cloud Gateway a first-class Kubernetes citizen and a truly enterprise-grade product focused on enterprise-grade problems. This is summed up in the following diagram:

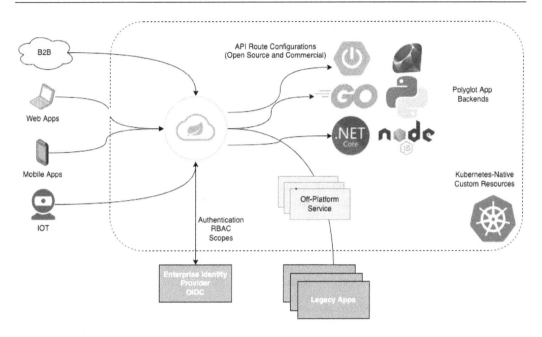

Figure 5.1 – Spring Cloud Gateway for Kubernetes at a glance

Now that we've covered how to deploy APIs, let's move on to how best to consume them. Let's have a look at **API Portal for VMware Tanzu**.

Why API Portal for VMware Tanzu?

Just as **Spring Cloud Gateway for Kubernetes** is a self-service tool allowing developers and operators to uniformly deploy, manage, and secure their APIs, **API Portal for VMware Tanzu** is a self-service tool that allows developers and consumers to publish, document, search for, discover, and try out those APIs.

Just as Spring Cloud Gateway for Kubernetes makes it easy to deploy multiple gateways for multiple use cases or environments, API Portal for Vmware Tanzu makes it easy to deploy multiple searchable catalogs across those very same use cases or environments.

This distributed, Kubernetes-native approach to API catalogs can scale across multiple large teams and enables self-service, which, in turn, gives you greater developer productivity and satisfaction.

Furthermore, it's a simple, straightforward operation to integrate API portals with enterprise SSO so that the right people can find the right APIs for their needs.

Now that we know *why* we're learning about these products, let's get started with installing them and trying them out. Let's go over some technical requirements; then, we can start with **Spring Cloud Gateway for Kubernetes**.

Technical requirements

API Portal for VMware Tanzu is a subcomponent of another product covered in this book, **Tanzu Application Platform (TAP)**. To get started with API Portal for VMware Tanzu, we'll first need to install a Kubernetes cluster and then install TAP on that cluster. You're free to use whichever Kubernetes distribution you prefer. If you'd like some guidance around standing up a cluster, you can go to the appendix at the end of the book where we describe several options for getting Kubernetes up and running.

Once you have a running Kubernetes cluster, you'll need to jump briefly to *Chapter 8*, *Enhancing Developer Productivity with Tanzu Application Platform*, and install *Tanzu Application Platform*. That chapter walks through a more complex end-to-end use case, while this chapter focuses solely on the *API Portal* component.

Next, you'll need an account on a container registry that presents a TLS certificate trusted by your Kubernetes cluster. The easiest thing to do is sign up for a free developer account at Docker Hub. Simply visit `https://hub.docker.com` and sign up.

If you want to deploy the real-world use case, you'll need an OIDC provider for SSO. If you don't already have something available, I'd recommend setting up a free developer account at Okta: `https://developer.okta.com/signup/`. Later in this chapter, there will be some detailed instructions for setting up your account to work with our real-world example application.

At the time of writing, the **Spring Cloud Gateway for Kubernetes** installation requires the *helm* CLI, which you can find here: `https://helm.sh/docs/intro/install/`.

Finally, we'll need the *kustomize* CLI, which can be found here: `https://kustomize.io`. If you just want to download the binary, you can also go here: `https://kubectl.docs.kubernetes.io/installation/kustomize/binaries`.

Now that we have Kubernetes running *Tanzu Application Platform*, we can install **Spring Cloud Gateway for Kubernetes** and **API Portal for VMware Tanzu**.

First, let's double-check that we have all our prerequisites in place:

- We need an up-to-date version of the `kapp` controller – at least 0.29.0:

  ```
  $ kubectl get deploy -n kapp-controller kapp-controller
  -ojsonpath='{.metadata.annotations.kapp-controller\.
  carvel\.dev/version}'
  v0.29.0
  ```

- We'll also need the `secretgen` controller:

  ```
  $ kubectl get deploy -n secretgen-controller
  NAME                      READY    UP-TO-
  ```

```
DATE    AVAILABLE    AGE
secretgen-
controller   1/1      1            1           23h
```

- Let's make sure we have up-to-date versions of `helm` and `kustomize` installed:

```
$ helm version
version.BuildInfo{Version:"v3.5.3",
GitCommit:"041ce5a2c17a58be0fcd5f5e16fb3e7e95fea622",
GitTreeState:"dirty", GoVersion:"go1.15.8"}


$ kustomize version
{Version:kustomize/v4.0.5
GitCommit:9e8e7a7fe99ec9fbf801463e8607928322fc5245
BuildDate:2021-03-08T20:53:03Z GoOs:darwin GoArch:amd64}
```

- And finally, we should test that we have a writable Docker repository and a working Docker CLI. The following is a simple way to test that everything has been set up properly. I'll use dockerhub as the registry, but you can plug in whichever solution you've chosen to use:

```
$ export DOCKER_USER=<your dockerhub username>
$ docker pull hello-world
$ docker tag hello-world docker.io/$DOCKER_USER/hello-
world
$ docker push docker.io/$DOCKER_USER/hello-world
```

If everything worked without errors, congratulations! You should have all the technical requirements in place to proceed to the next section, where you will install **Spring Cloud Gateway for Kubernetes** and **API Portal for VMware Tanzu**!

Spring Cloud Gateway for Kubernetes – getting started

Let's start by installing **Spring Cloud Gateway for Kubernetes**. There are two distinct parts to this product: the Kubernetes Operator and the deployed gateways.

The *Kubernetes Operator* works with the Kubernetes API to handle three *Custom Resource Definitions*:

- **SpringCloudGateway** (abbr: *scg*): This is an (optionally) HA instance of Spring Cloud Gateway – centrally packaged, deployed, and life cycle-managed for the developer.

- **SpringCloudGatewayRouteConfig** (abbr: *scgrc*): This is a set of instructions for a Spring Cloud Gateway instance around what host/path to accept requests on, how to filter that request, and which backend Kubernetes service to eventually proxy it onto.

- **SpringCloudGatewayMapping** (abbr: *scgm*): This tells the Kubernetes operator which *SpringCloudGatewayRouteConfigs* are associated with which *SpringCloudGateways*. The operator automatically reconfigures *SpringCloudGateways* with the necessary Spring configuration properties to match the contents of *SpringCloudGatewayRouteConfig*.

The deployed instances of Spring Cloud Gateway are the other half of the product. The Kubernetes operator stands up multiples of these gateways, which can accept incoming traffic via a Kubernetes service or ingress. It then looks for *SpringCloudGatewayRouteConfigs* that are mapped to a particular Spring Cloud Gateway instance and dynamically configures those routes.

The Spring Cloud Gateway deployed instances are what handle API traffic for developers' APIs. The Kubernetes Operator is how developers can self-service provision Spring Cloud Gateway instances and (human, not Kubernetes) operators can monitor and manage those instances automatically, and at scale.

The installation process is as simple as running a couple of scripts to relocate the necessary container images to your container registry and then deploying the necessary Kubernetes artifacts to your cluster using *Helm*. The official installation instructions can be found here: `https://docs.vmware.com/en/VMware-Spring-Cloud-Gateway-for-Kubernetes/1.0/scg-k8s/GUID-installation.html`.

To paraphrase the installation, here are the steps:

1. Download the binary from this link: `https://network.tanzu.vmware.com/products/spring-cloud-gateway-for-kubernetes`.
2. Unzip it.
3. Run the script to take the local `.tgz` image layers and push them to your container repository.
4. Run the script to deploy everything to your Kubernetes cluster via *helm*.

First, let's relocate our images. The Kubernetes operator and the deployed Spring Cloud Gateway instances need to exist in an external image repository that is reachable from the Kubernetes cluster. At this time, Spring Cloud Gateway for Kubernetes is delivered via download and all the image bits are contained in the downloaded file. To get those bits somewhere Kubernetes can access them, we'll need to load them into our local Docker environment and push them to the remote repository. The straightforward Docker commands are in the `scripts/relocate-images.sh` file if you're interested. Otherwise, simply navigate to the `scripts` directory and call the script. I'll share the abbreviated output from my run. `docker.io/rhardt` is my repository. If you set up a Docker Hub account, yours will be `docker.io/<your-username>`:

```
bash-5.0$ ./relocate-images.sh docker.io/rhardt
Relocating image
=================
image name: gateway
version: 1.0.8
```

```
source repository: registry.tanzu.vmware.com/spring-cloud-
gateway-for-kubernetes/gateway:1.0.8
destination repository: docker.io/rhardt/gateway:1.0.8
Loaded image: registry.tanzu.vmware.com/spring-cloud-gateway-
for-kubernetes/gateway:1.0.8
The push refers to repository [docker.io/rhardt/gateway]
1dc94a70dbaa: Layer already exists
… (more layers)
824bf068fd3d: Layer already exists
1.0.8: digest: sha256:8c1deade58dddad7ef1ca6928cbdd76e401bc0a-
faf7c44378d296bf3c7474838 size: 4500
Relocating image
=================
image name: scg-operator
version: 1.0.8
source repository: registry.tanzu.vmware.com/spring-cloud-gate-
way-for-kubernetes/scg-operator:1.0.8
destination repository: docker.io/rhardt/scg-operator:1.0.8
Loaded image: registry.tanzu.vmware.com/spring-cloud-gateway-
for-kubernetes/scg-operator:1.0.8
The push refers to repository [docker.io/rhardt/scg-operator]
1dc94a70dbaa: Layer already exists
… (more layers)
824bf068fd3d: Layer already exists
1.0.8: digest: sha256:da9f2677e437ccd8d793427e6cafd9f4b-
b6287ecffdc40773cf3b1f518f075fb size: 4498
```

So, now, we have our container images somewhere that Kubernetes can find them. The other thing that the `relocate-images.sh` script does is create a file called `scg-image-values.yaml` that Helm will use when deploying to Kubernetes.

The second script we will run is `install-spring-cloud-gateway.sh`, also in the `scripts` directory. I encourage you to peruse this script as well as it's an excellent example of a comprehensive helm deployment, complete with updating an existing installation, specific instructions on timeout, error diagnostics in the event of failure, and checking to see that the deployment succeeded:

```
bash-5.0$ ./install-spring-cloud-gateway.sh
chart tarball: spring-cloud-gateway-1.0.8.tgz
chart name: spring-cloud-gateway
Waiting up to 2m for helm installation to complete
```

```
Release "spring-cloud-gateway" does not exist. Installing it
now.
NAME: spring-cloud-gateway
LAST DEPLOYED: Mon Feb 21 18:25:13 2022
NAMESPACE: spring-cloud-gateway
STATUS: deployed
REVISION: 1
TEST SUITE: None
NOTES:
This chart contains the Kubernetes operator for Spring Cloud
Gateway.
Install the chart spring-cloud-gateway-crds before installing
this chart

Checking Operator pod state
deployment "scg-operator" successfully rolled out
↪ Operator pods are running

Checking custom resource definitions
↪ springcloudgatewaymappings.tanzu.vmware.com successfully
installed
↪ springcloudgatewayrouteconfigs.tanzu.vmware.com successfully
installed
↪ springcloudgateways.tanzu.vmware.com successfully installed

Successfully installed Spring Cloud Gateway operator
bash-5.0$
```

Now that we've installed Spring Cloud Gateway for Kubernetes, let's move on to API Portal for VMware Tanzu.

API Portal for VMware Tanzu – getting started

Unlike *Spring Cloud Gateway for Kubernetes*, API Portal for VMware Tanzu comes as a part of *Tanzu Application Platform*, which you installed in the *Technical requirements* section previously. You can verify that API Portal is installed and running by navigating to your Tanzu Application GUI, either at a local port forward address such as `http://localhost:8081` or at your ingress domain such as `http://tap-gui.example.com`. This depends on your choice of Kubernetes cluster and how you installed TAP. Once you access your Tanzu Application Platform GUI, click the **APIs** menu item on the left-hand side; you should see a screen like the following:

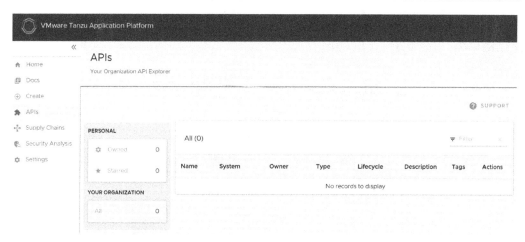

Figure 5.2 – API Portal

Spring Cloud Gateway for Kubernetes and API Portal for VMware Tanzu – real-world use case

Now, we're going to implement a real-world use case using **API Portal for VMware Tanzu** and **Spring Cloud Gateway for Kubernetes**. This will show how these products naturally work together and how, in combination, you can realize significant value as a developer, consumer, or operator of APIs.

Here's a breakdown of our next steps:

1. We will download an API-driven application called *Animal Rescue* while approximating a real-world animal adoption site, complete with SSO.

2. We will set up our Okta environment so that our application can use Okta for SSO. Then, we will take the Okta parameters and plug them into the Animal Rescue application.

3. We will deploy the Animal Rescue application to our Kubernetes cluster, which will create the necessary Spring Cloud Gateway artifacts. By doing so, we will be able to browse and interact with the Animal Rescue web app.

4. We will modify our API Portal installation so that it connects to the Spring Cloud Gateway Kubernetes Operator, which now knows all the API routes for the Animal Rescue app and exposes them as OpenAPIv3.

5. We will port-forward to our API Portal to explore and try out the Animal Rescue APIs.

Now that you know what we're trying to accomplish, here are the steps.

Step 1 – cloning this book's Git repository and navigating to the Animal Rescue application

Use the Git CLI to clone the book code repository and then navigate to the *animal-rescue* application as seen in the following code snippet:

```
bash-3.2$ git clone https://github.com/PacktPublishing/
DevSecOps-in-Practice-with-VMware-Tanzu
Cloning into 'DevSecOps-in-Practice-with-VMware-Tanzu'...
remote: Enumerating objects: 499, done.
remote: Counting objects: 100% (499/499), done.
remote: Compressing objects: 100% (351/351), done.
remote: Total 499 (delta 95), reused 473 (delta 75), pack-
reused 0
Receiving objects: 100% (499/499), 9.80 MiB | 10.39 MiB/s,
done.
Resolving deltas: 100% (95/95), done.
bash-3.2$ cd DevSecOps-in-Practice-with-VMware-Tanzu/chapter-
05/animal-rescue/
bash-3.2$
```

Step 2 – configuring your Okta developer account to provide SSO for Animal Rescue

These steps are a bit involved, so I recommend following the instructions directly from the VMware documentation here: https://docs.vmware.com/en/VMware-Spring-Cloud-Gateway-for-Kubernetes/1.0/scg-k8s/GUID-sso-setup-guide.html.

After completing the setup, we'll need to complete one additional step for this example to work. We need to add two additional sign-in redirect URIs for the "gateway" application. This screenshot shows my Okta setup:

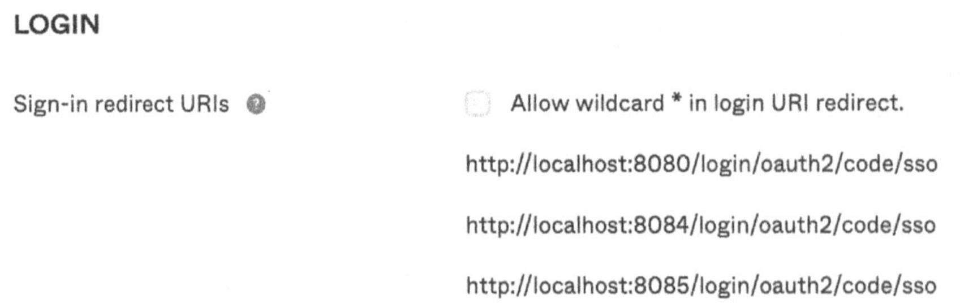

LOGIN

Sign-in redirect URIs ❓   ☐   Allow wildcard * in login URI redirect.

                                            http://localhost:8080/login/oauth2/code/sso

                                            http://localhost:8084/login/oauth2/code/sso

                                            http://localhost:8085/login/oauth2/code/sso

Figure 5.3 – Screenshot of the Okta sign-in redirect URIs

We'll be port-forwarding the Animal Rescue app to `http://localhost:8084` and the API gateway to `http://localhost:8085`; we'd like both to be able to authenticate via Okta.

At the end of this exercise, you'll need three pieces of information from your Okta environment: `issuer-id`, `client-id`, and `client-secret`.

The VMware documentation instructs you on how to configure the Animal Rescue app with that information. I'll include it here as well. These steps are relative to the `animal-rescue` app that we cloned from GitHub in the previous section:

- Create `backend/secrets/sso-credentials.txt` with the following:

  ```
  jwk-set-uri=<issuer uri>/v1/keys
  ```

- Create `gateway/sso-secret-for-gateway/secrets/test-sso-credentials.txt` with the following:

  ```
  scope=openid,profile,email,groups,animals.adopt
  client-id=<client id>
  client-secret=<client id>
  issuer-uri=<issuer uri>
  ```

Some of the additional steps mentioned in the documentation, such as editing `roles-attribute-name` or configuring route security have already been done for you in the version of the app you downloaded.

Step 3 – deploying the Animal Rescue application

From the `animal-rescue` directory – that is, the home directory of the app you cloned (`DevSecOps-in-Practice-with-VMware-Tanzu/chapter-05/animal-rescue`) – run the following command:

```
bash-5.0$ kustomize build . | kubectl apply -f -
namespace/animal-rescue created
secret/animal-rescue-sso created
secret/sso-credentials created
service/animal-rescue-backend created
service/animal-rescue-frontend created
deployment.apps/animal-rescue-backend created
deployment.apps/animal-rescue-frontend created
ingress.networking.k8s.io/gateway-demo created
springcloudgateway.tanzu.vmware.com/gateway-demo created
```

```
springcloudgatewaymapping.tanzu.vmware.com/animal-rescue-
backend-routes created
springcloudgatewaymapping.tanzu.vmware.com/animal-rescue-
frontend-routes created
springcloudgatewayrouteconfig.tanzu.vmware.com/animal-rescue-
backend-route-config created
springcloudgatewayrouteconfig.tanzu.vmware.com/animal-rescue-
frontend-route-config created
```

At this point, you can watch to see when all the pods in the `animal-rescue` namespace have come up:

```
bash-5.0$ kubectl get pods -n animal-rescue -w
NAME                                     READY    STATUS    RES
TARTS    AGE
animal-rescue-backend-546fc6c569-
kgj2s    1/1       Running    0           27m
animal-rescue-frontend-
b74f54847-rq284    1/1       Running    0            27m
gate-
way-demo-0                    1/1       Running    0            23m
gate-
way-demo-1                    1/1       Running    0            24m
```

Once the pods are up, we can port forward to `localhost:8084`.

At this point, I highly recommend an *incognito* browser window to prevent the application from getting confused regarding your Okta developer account.

Open your browser to `http://localhost:8084`; you should see the Animal Rescue web application replete with cute animals to adopt and an SSO login button in the top-right corner. If you configured everything correctly, clicking that button will take you to an Okta login and then back to the running application once you've logged in. This is what it looks like in my browser. Notice the text in the top right with my Okta username:

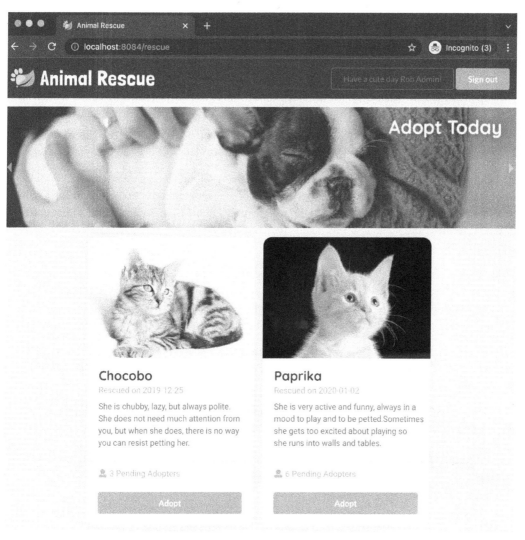

Figure 5.4 – Animal Rescue screenshot

So far, so good! Now, let's take things to the next level and integrate this API with **API Portal for VMware Tanzu**.

Step 4 – (optional) viewing the Animal Rescue API's OpenAPI 3.0 specification

This is, in my opinion, one of the most compelling features of this pair of products.

In practice, developers are responsible for manually keeping their API specification in some sort of documentation format, usually OpenAPI 3.x or Swagger. This is tedious for the developer, and in the heat of battle, that API specification will be the first thing to stop being maintained. This means that new consumers will be basing their clients on incorrect, out-of-date API documentation.

Once you deploy the *Animal Rescue* application, the Spring Cloud Gateway Kubernetes Operator manages all the routes – that is, the Kubernetes Operator manages the configuration of Spring Cloud Gateway instances so that every API call gets properly routed to the right backend service, along with any filters, security, and metadata. This is sufficient information to auto-generate the API's OpenAPI 3.0 specification.

It's hard to understate what a breakthrough this is. The actual Kubernetes objects that control the API's runtime configuration are used to auto-generate always-accurate, always-up-to-date OpenAPI 3.0 documentation. This removes the burden of documentation creation and maintenance from the API developer, and the burden of working around out-of-date documentation from the API consumer.

Let's take a quick look at our Animal Rescue API auto-generated documentation. The OpenAPI 3.0 documentation for every deployed API across all gateways is available via a web endpoint on the Spring Cloud Gateway Operator. That web endpoint is exposed via a ClusterIP service.

In production, you may want to put a TLS-secured ingress in front of that service, but for the sake of demonstration, let's just set up a quick port forward:

```
bash-5.0$ kubectl get svc -n spring-cloud-gateway
NAME            TYPE         CLUSTER-IP       EXTER-
NAL-IP    PORT(S)    AGE
scg-operator    ClusterIP    10.98.103.232    <none>         80/
TCP      40h
bash-5.0$ kubectl port-forward -n spring-cloud-gateway svc/
scg-operator 8083:80
Forwarding from 127.0.0.1:8083 -> 8080
Forwarding from [::1]:8083 -> 8080
```

First, we determined that the `spring-cloud-gateway` service was running on port 80, so we port-forwarded our local workstation's port 8083 to that service on port 80. Now, we can point a web browser to `http://localhost:8083`, and voila! You will see the OpenAPI 3.0.1 representation of the `Animal Rescue` API, as shown in the following screenshot:

```
- {
      openapi: "3.0.1",
    - info: {
          title: "Animal Rescue",
          description: "Sample application for Spring Cloud Gateway commercial product demos.",
          version: "1.0.0-K8s"
      },
    - externalDocs: {
          url: "https://github.com/spring-cloud-services-samples/animal-rescue/"
      },
    - servers: [
        - {
              url: "http://animal-rescue.my.domain.io"
          }
      ],
    - paths: {
        - /api/animals: {
              summary: "Route ID: animal-rescue-animal-rescue-backend-routes-0",
            - get: {
                - tags: [
                      "pet adoption"
                  ],
                  summary: "Retrieve pets for adoption.",
                  description: "Retrieve all of the animals who are up for pet adoption.",
                - responses: {
                    - 200: {
                          description: "Ok",
                        - headers: {
                            - X-Remaining: {
                                  description: "RateLimit: number of requests remaining",
                                - schema: {
                                      type: "integer"
                                  }
                              }
                          }
                      },
                    - 429: {
                          description: "Too Many Requests. RateLimit=2,10s",
                        - headers: {
                            - X-Retry-In: {
                                  description: "RateLimit: time in milliseconds until retry",
                                - schema: {
                                      type: "integer"
                                  }
                              }
                          }
                      }
```

0].paths["/api/animals"].get.responses["429"]

Figure 5.5 – Animal Rescue auto-generated OpenAPI 3.0.1 specification

This is a 100% valid and accurate OpenAPI 3.0, so any tools that can generate clients or server stubs from that format will work with it. Furthermore, the Kubernetes `SpringCloudGatewayRouteConfig` objects contain fields for the following:

- Human-readable descriptions
- All the possible response codes and their meanings
- Any sort of authentication required

In other words, the real-world Kubernetes objects driving the real-world runtime behavior are also driving the documentation, so it's always accurate and up to date.

Step 5 – (optional) connecting API Portal for VMware Tanzu with the Spring Cloud Gateway for Kubernetes OpenAPI endpoint

As we demonstrated earlier, **API Portal for VMware Tanzu** is driven entirely by OpenAPI 3 or Swagger documents. It's only logical, then, that we'd set up the API Portal to consume the OpenAPI 3 documents generated by **Spring Cloud Gateway for Kubernetes**. If you set up a port forward in the previous step, you can exit from it. Since the API Portal and Spring Cloud Gateway are on the same Kubernetes cluster, they can communicate via ClusterIP services.

We previously installed **API Portal for VMware Tanzu** using the `tanzu package` command. Now, let's throw together a simple configuration file that overrides where API Portal will look for OpenAPI 3 docs, and update the package installation.

First, create a file called `api-portal-values.yaml` with the following contents:

```
apiPortalServer:
    sourceUrls: https://petstore.swagger.io/v2/swagger.
json,https://petstore3.swagger.io/api/v3/openapi.json,http://
scg-operator.spring-cloud-gateway.svc.cluster.local/openapi
```

As you can see, we're appending the ClusterIP service of the Spring Cloud Gateway Operator to the list of places where the API Portal will search for API documentation.

Now, we can use the `tanzu package` command to update the installation. Behind the scenes, the API Portal will be redeployed with the new configuration. This method of deploying software removes configuration drift as any manual configuration changes to the deployment will be overwritten by the `kapp` controller:

```
bash-5.0$ tanzu package installed update api-portal -n
tap-install -f api-portal-values.yaml
| Updating installed package 'api-portal'
| Getting package install for 'api-portal'
```

```
| Getting package metadata for 'api-portal.tanzu.vmware.com'
| Creating secret 'api-portal-tap-install-values'
| Updating package install for 'api-portal'
/ Waiting for 'PackageInstall' reconciliation for 'api-portal'
/ 'PackageInstall' resource install status: Reconciling

Updated installed package 'api-portal' in namespace
'tap-install'
bash-5.0$
```

Now, let's re-instate our port forward to our API Portal:

```
bash-5.0$ kubectl get svc -n api-portal
NAME                TYPE          CLUSTER-IP        EXTER-
NAL-IP    PORT(S)    AGE
api-portal-server   Clus-
terIP    10.103.142.154    <none>           8080/TCP    40h
bash-5.0$ kubectl port-forward -n api-portal svc/api-por-
tal-server 8085:8080
Forwarding from 127.0.0.1:8085 -> 8080
```

And upon opening a web browser to `http://localhost:8085`, we should see our `Animal Rescue` API in the list, as shown in the following screenshot:

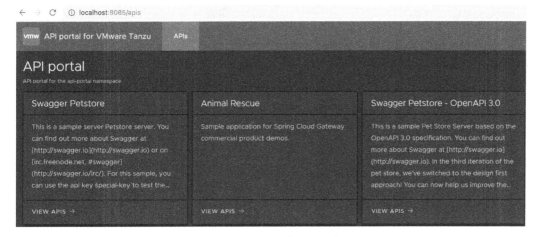

Figure 5.6 – API Portal with the Animal Rescue API

In the `Animal Rescue` tile, click **VIEW APIS**. Then, on the resulting screen, click **/api/animals**, then **Try it Out**, then **Execute**. If everything has been configured correctly, you should see some animal data in JSON format in the **Server response** section, as shown in the following screenshot:

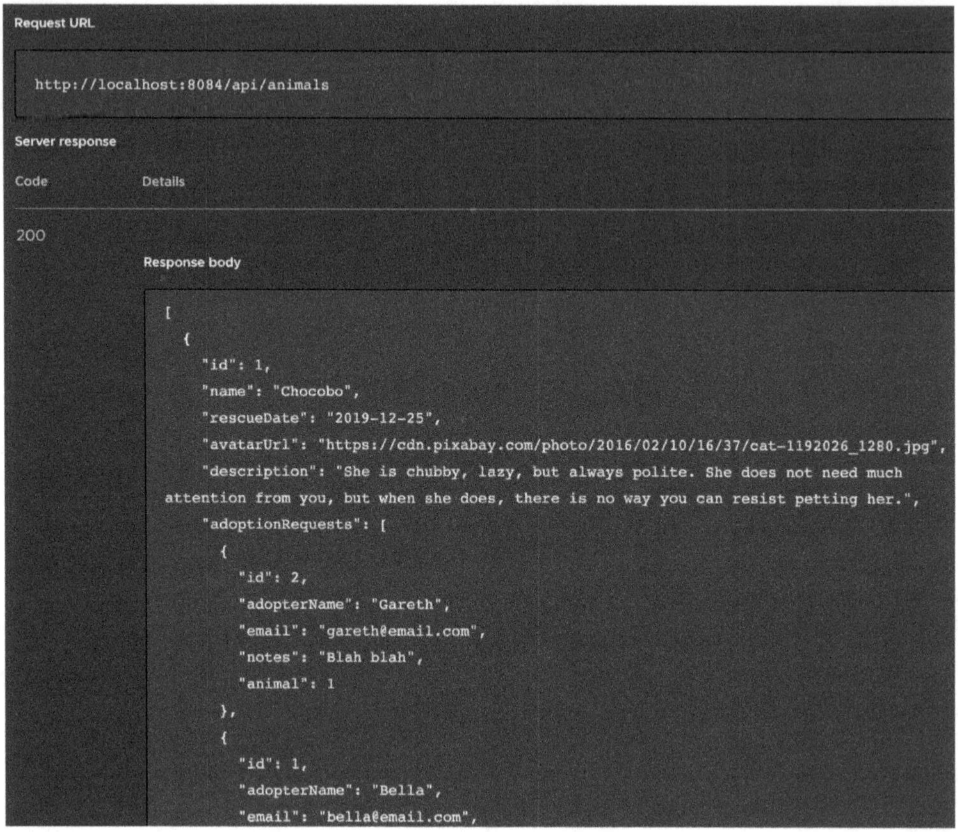

```
Request URL

  http://localhost:8084/api/animals

Server response

Code      Details

200
          Response body

          [
            {
              "id": 1,
              "name": "Chocobo",
              "rescueDate": "2019-12-25",
              "avatarUrl": "https://cdn.pixabay.com/photo/2016/02/10/16/37/cat-1192026_1280.jpg",
              "description": "She is chubby, lazy, but always polite. She does not need much
          attention from you, but when she does, there is no way you can resist petting her.",
              "adoptionRequests": [
                {
                  "id": 2,
                  "adopterName": "Gareth",
                  "email": "gareth@email.com",
                  "notes": "Blah blah",
                  "animal": 1
                },
                {
                  "id": 1,
                  "adopterName": "Bella",
                  "email": "bella@email.com",
```

Figure 5.7 – Trying out APIs from the browser

Just to be clear, you're able to *try out* APIs directly from the API Portal interface.

After the exhilarating experience of installing these two products, wiring them together, and implementing a complex real-world application, we must face the inevitable. Long after the fun of deploying a working piece of software has faded, the ongoing task of maintenance and support will remain. We'll cover that next.

Common day-2 operations with Tanzu Application Accelerator

Congratulations! If you've made it this far, you have set up an API publishing and consumption model that can unlock incredible value in your enterprise software development operation.

As is usually the case in the enterprise, the hard problem is not getting something installed and configured, it's operating it over time. Let's look at some of these day-2 operations.

Updating an API's route definition

Let's say we identify a typo in an API route's description. We'd like to be able to update that route in real time and have it propagate to the API Portal without any additional configuration. Let's do that. Navigate to the `animal-rescue` app that you cloned from GitHub and open the `./backend/k8s/animal-rescue-backend-route-config.yaml` file in an editor. At line 17, you'll see a description starting with *Retrieve all the anmals*. Fix it so that it reads *Retrieve all the animals*. If you want to be doubly sure, insert some additional text if you'd like. Here's what it looked like in my editor. I wanted to be *especially* sure that I was seeing my changes:

```
      - "pet adoption"
    title: "Retrieve pets for adoption."
    description: "ROB EDIT - Retrieve all of the anmals who are up for pet
- predicates:
    - Path=/api/animals/{animalId}/adoption-requests
    - Method=POST
  ssoEnabled: true
  tokenRelay: true
```

Figure 5.8 – Editing the description of an API route

Now, we must simply update the Spring Cloud Gateway Route Config in Kubernetes, and the documentation should auto-update. Don't forget the namespace, which in the original install was added by *kustomize*:

```
$ kubectl apply -f ./animal-rescue-backend-route-config.yaml -n
animal-rescue

springcloudgatewayrouteconfig.tanzu.vmware.com/animal-rescue-
backend-route-config configured
```

Now, if you still have the port-forward to `api-portal` running, you can navigate to `http://localhost:8085`, click through to the Animal Rescue `/api/animals` route, and see your update in real time!

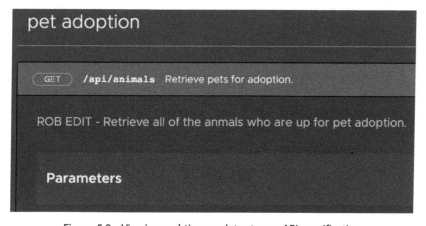

Figure 5.9 – Viewing real-time updates to our API specification

At this point, we've gone in-depth into numerous *day-0* and *day-1* tasks, such as deploying and consuming APIs. But anyone who has spent any time at all in the enterprise software space knows that the vast majority of our time and effort is spent on the *day-2* tasks: updating, monitoring, measuring, and improving our software. Let's briefly cover some of these day-2 topics.

Updating the API Portal package

The `tanzu package` commands in conjunction with the kapp controller give you incredible power to life cycle-manage your deployed software. In the case of API Portal, you can grab the currently deployed version and see which versions are currently available. Then, to update it, you simply execute `tanzu package installed update -n tap-install api-portal -v <new version>`:

```
bash-5.0$ tanzu package installed list -n tap-install
| Retrieving installed packages...
  NAME          PACKAGE-NAME                      PACKAGE-
VERSION   STATUS
  api-portal    api-portal.tanzu.vmware.
com   1.0.8                 Reconcile succeeded
```

We currently have 1.0.8 installed:

```
bash-5.0$ tanzu package available list -n tap-install
api-portal.tanzu.vmware.com
/ Retrieving package versions for api-portal.tanzu.vmware.
com...
  NAME                          VERSION   RELEASED-AT
  api-portal.tanzu.vmware.com   1.0.8     2021-12-15 19:00:00
-0500 EST
  api-portal.tanzu.vmware.com   1.0.9     2022-01-02 19:00:00
-0500 EST
```

It appears that version 1.0.9 became available recently:

```
bash-5.0$ tanzu package installed update api-portal -n
tap-install -v 1.0.9
| Updating installed package 'api-portal'
| Getting package install for 'api-portal'
| Getting package metadata for 'api-portal.tanzu.vmware.com'
| Updating package install for 'api-portal'
```

```
/ Waiting for 'PackageInstall' reconciliation for 'api-portal'


Updated installed package 'api-portal' in namespace
'tap-install'
```

And with a simple command, our `api-portal` package auto-updates to the latest version.

Summary

APIs are everywhere. With OpenAPI 3 and REST becoming the well-accepted standard among developers, a significant portion of an enterprise developer's job consists of creating, publishing, discovering, and consuming APIs.

In an environment with a significant number of developers on disparate teams, this inevitably leads to waste: a significant waste of time, effort, and money.

Developers waste energy reverse-engineering inadequately documented APIs. They waste time identifying the right API. They burn unnecessary cycles figuring out how to make their API widely available.

Once they have an API deployed, there's a waste of time and energy making sure the API is adequately secured and that it's shared across a large organization, not to mention making sure those APIs are maintained and kept up to date.

Architects may spend undue time building out a central clearing house for discovering APIs and a central gateway for deploying them.

Finally, operators are tasked with monitoring and measuring many polyglot APIs across different platforms.

With the tools that you've now become familiar with in this chapter, I hope you can take a big bite out of that API waste, making your team, your development org, and your company wildly successful, efficient, and API-driven.

With our next chapter, we will move our focus from *building* software to *running* it. We'll kick off our new area of focus with a deep dive into hosting and maintaining OCI container images and Helm Charts with the **VMware Harbor Registry**.

Part 2 – Running Cloud-Native Applications on Tanzu

This part of the book focuses on the tools and techniques that help run modern cloud-native applications on Kubernetes with multi-cloud and multi-cluster deployments.

This part of the book comprises the following chapters:

- *Chapter 6, Managing Container Images with Harbor*
- *Chapter 7, Orchestrating Containers across Clouds with Tanzu Kubernetes Grid*
- *Chapter 8, Enhancing Developer Productivity with Tanzu Application Platform*

6

Managing Container Images with Harbor

In the previous chapters, we covered the tools in the Tanzu portfolio that help us build cloud-native applications. We started our first segment with an overview of the evolution of building, running, and managing modern cloud-native applications and their platforms. Then, we saw how we can start application development using templates, how to build secure container images, how to quickly provision backing services for the applications, and how to manage APIs using various Tanzu products. After learning about building cloud-native applications, in this chapter, we will take a deep dive into various aspects of running them.

As the title of this chapter indicates, we will learn how to manage our container images and securely make them accessible using Harbor to deploy our applications on Kubernetes. Harbor is an open source container registry project under the **Cloud Native Computing Foundation** (**CNCF**) umbrella. Despite Harbor being a fully open source tool, we have included it in this book for three main reasons. Firstly, Harbor was incubated by VMware and donated to CNCF in mid-2018. VMware is also one of the major contributors to the project and has actively invested in Harbor since then. Secondly, Harbor has also been recognized as a graduate project under the CNCF umbrella, which is a state that is tagged as a very popular, mature, and stable project within the CNCF ecosystem. Finally, the main reason to include Harbor in this book is that VMware, being a significant stakeholder in this project, also provides commercial enterprise support for Harbor as a part of its Tanzu portfolio.

> **Sidenote**
> Henceforth, in this chapter, we will refer to a *container image* as an *image* only for brevity.

In this chapter, we will cover Harbor in detail by covering the following topics:

- **Why Harbor?**: A walkthrough of the features and capabilities of Harbor

- **Unboxing Harbor**: A detailed overview of the anatomy of Harbor

- **Getting started with Harbor**: Learn how to install and configure Harbor

- **Common day-2 operations with Harbor**: Learn how to perform various configuration and usage-related activities on Harbor

Let's start by learning about the background of Harbor.

Why Harbor?

In this section, we will review the various features, capabilities, and reasons to consider using Harbor as a container registry. These reasons will be explained using the security, control, and extensibility features of Harbor as described henceforth.

Using Harbor for security

There are some strong security reasons and features that make Harbor a good choice for a container registry, which shifts security to a proactive measure rather than reactive in the applications' journey toward production. Let's review these security benefits:

- Harbor comes with the capability to scan each image for the presence of **critical vulnerability exposures (CVEs)** as a result of certain software libraries and operating system versions used in the image. Such scanning provides a detailed report of the CVEs found in the corresponding image, along with their severity level, details of the exposure, and the version of the software in which that CVE is remediated. We can get such scanning results using either the web portal or using the REST APIs provided by Harbor. Harbor also allows you to use an external image scanning tool in place of or in addition to the default one. Such visibility of the possible security loopholes in the images could provide a preventative security posture well in advance in the application deployment process.

- Depending on the application environment and the preferred tolerance level, Harbor also provides a way to prevent its clients from pulling such images that are scanned for CVEs and contain CVEs higher than the allowed severity level. For example, we can configure a policy in Harbor that any image that has CVEs found with categories more than medium severity in a project named `Production Repo` may not be pulled to deploy containers. This capability provides required guardrails to prevent damage at the front gate itself. It ensures that the flagged images are never allowed to be pulled to run workloads and allow bad actors to exploit them later.

- Harbor also supports integrations with Notary (`https://github.com/notaryproject/notary`), which is an open source project that can digitally sign the images for authenticity. You can create a container deployment pipeline using such an image signing utility to allow only signed and hence authorized images to be deployed in your production environment. Such an arrangement can greatly enhance your security posture as no unverified, unscanned, or potentially dangerous images can be deployed in your environment.

- Harbor has robust **role-based access control** (**RBAC**) capabilities. It allows you to configure users with two levels, mainly at the project level and at the system level, to provide the required control and flexibility for a multi-tenant environment.

  Moreover, Harbor also allows you to separate user accounts from system accounts (known as **robot accounts** in Harbor) that can be used for **continuous integration** (**CI**) and **continuous deployment** (**CD**) automation processes. We may specify required permissions to such robot accounts to perform only allowed operations using the automation processes.

- We may create a hub-and-spoke architecture while using Harbor as the hub that replicates images to and from either external or other internal Harbor container registries. An example of such a deployment is shown in *Figure 6.1*. Such an arrangement may allow organizations to prevent their internal users from pulling arbitrary and insecure images from unauthorized sources. But at the same time, it allows them to pull those images from only the internally deployed Harbor, which would have replicated authorized images from an external image repository. This feature provides the required control to ensure security without affecting developers' freedom and productivity.

- As we will see later in this chapter, Harbor has several components and supports many external integrations for various capabilities. To ensure the safety of such data transfers, all these inter-component communication channels use **Transport Layer Security** (**TLS**) encryption.

After reviewing the key features of Harbor around security, let's check what its benefits are from an operational control point of view.

Using Harbor for operational control

There are several popular container registries available in the market as online **Software-as-a-Service** (**SaaS**) offerings, including Docker Hub, **Google Container Registry** (**GCR**), and offerings from many other cloud providers. The point where Harbor differs from these online options is the fact that it can be deployed in an air-gapped environment. When there is a need to keep the application images private and on-premise, we need an offering like Harbor. With such on-premises deployments, as discussed in the previous section about security-specific reasons, Harbor provides a control mechanism to expose only authorized images that are replicated in Harbor from external sources for internal consumption. This way, the operators can prevent internal image users from downloading potentially vulnerable images from unauthorized sources.

Additionally, Harbor is also an open source community-driven project that is at the **Graduated** maturity level in CNCF, like Kubernetes. CNCF only graduates an open source project when there is a significant community contribution and adoption. Since VMware is one of the major contributors to the project, it also provides commercial support for Harbor.

Along with the point of being a CNCF-mature and commercially supported open source project, Harbor has an array of multi-tenancy features. We will visit some of these features later in this chapter. But at a high level, Harbor admins can configure team-wise storage quotas for images and choose from different image vulnerability scanners, image retention periods, team-wise webhook configurations to trigger a CD pipeline, CVE whitelisting, and a few others. Having these configurations separate for different teams using the same deployment of Harbor provides the required operational control to Harbor admins, along with the required flexibility to the user groups.

Lastly, under the operational control area, Harbor provides various general administrative configurations that are common for the deployment and all user groups. Such configurations include the following administrative controls:

- Cleaning up all untagged artifacts using a garbage collection routine that can be triggered on-demand or scheduled
- Managing user groups and their permissions
- Configuring external or internal authentication providers, including **Light-weight Directory Access Protocol (LDAP)** and **Open ID Connect (OIDC)** systems
- Configuring custom **Open Container Initiative (OCI)** artifacts to store binary objects other than images in Harbor
- Configuring image proxy caching to allow externally hosted images to be stored in an offline mode
- Accessing key performance metrics to check on Harbor's health
- Enabling the distributed tracing telemetry data for enhanced troubleshooting capabilities for Harbor

> Tip
> **Open Container Initiative (OCI)** is an open governance structure for creating open industry standards around container formats and runtimes. Source: `https://opencontainers.org/`.

After seeing how Harbor can help to obtain control over various types of configurations, let's see one more category of reasons to use Harbor – its extensibility.

Using Harbor for its extensibility

Harbor is a solution that is comprised of a few different microservices that work together to serve the purpose of being a purpose-built container registry. It has several components that can be replaced with other available options providing similar functionalities. Moreover, we can extend some functionalities to provide more choices for the end users to pick from. Such areas of extensibility include integration with an external container registry, CVE scanners, authentication providers, and OCI-compliant objects that can be hosted on Harbor. The following sections describe them in detail.

Extending image sources and destinations through replication

Harbor allows you to create image replication rules to extend the image library using an external image repository. That way, the clients of Harbor can the pull required images from an external repository such as Docker Hub without accessing Docker Hub. Such extensions are helpful for an air-gapped deployment where open internet access and open image downloading from a public repository are not desirable from a security point of view. Additionally, Harbor allows you to create replication rules for push and pull operations for a bidirectional flow of artifacts. *Figure 6.1* shows how *the central Harbor repository* pulls (replicates) images from *Docker Hub* and *GCR* and then pushes those images to the *remote Harbor repositories* for a better network co-location for the nearby Kubernetes clusters. The arrows in the figure indicate the flow of images:

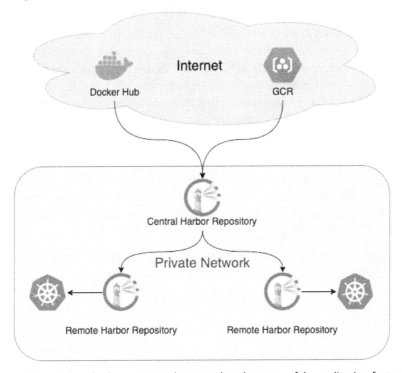

Figure 6.1 – Harbor deployment topology to take advantage of the replication feature

Such extensions of image repository locations for different sources and destinations can be very useful to provide controlled access to the replicated images from security and governance. Additionally, it can also help reduce network latency and bandwidth requirements.

In addition to using Harbor for replication, we can also configure Harbor as a proxy to cache externally located images. This caching arrangement can help save network latency in transferring frequently used images and save the network bandwidth required for internet traffic. Additionally, using Harbor for caching may cache only used images for a given timeframe. And if the image is not actively pulled, then it is removed. However, such a proxy configuration allows more freedom to access any available images versus only replicated ones. Both replication and caching have their use cases, pros, and cons.

Adding or replacing vulnerability scanners

By default, Harbor comes with Trivy (`https://github.com/aquasecurity/trivy`) for CVE scanning of images. However, you can incorporate your own instance of a Trivy implementation if you have one. You may also integrate a different CVE scanner with Harbor in place of or in addition to Trivy. This extension allows different teams to use their preferred scanner from the list of supported ones by Harbor. In the present scenario, Harbor supports Clair, Anchore, Aqua, DoSec, Sysdig Secure, and Tensor Security in addition to Trivy.

Extending authentication providers

Harbor provides database-level authentication where user accounts can be directly configured in Harbor as the default and primitive approach. However, the administrator may configure Harbor to use either an LDAP/Active Directory service or an OIDC provider. In that case, such external authentication providers will be used to create and manage user accounts. Harbor will redirect authentication requests to these external authentication providers and based on the identity provided by the authentication provider, Harbor grants the required access to the user.

Extending user-defined OCI artifacts hosting

Along with images, Harbor can also store Helm charts and other user-defined OCI artifacts. Such artifacts can be **Kubeflow** data models, which are used for machine learning on Kubernetes. For such extensions, the objects must follow Harbor-specific configuration using a manifest file. The use cases of such user-defined extensions are rare but possible.

So far in this chapter, we have seen different security, operational, and extensibility reasons explaining the *Why* behind using Harbor as a container repository. It is open source but supported by VMware and a lightweight, flexible, and purpose-built container registry that also helps enhance the overall container security posture via image scanning, replication, and signing features. In the next section of this chapter, we will discuss the *What* part of Harbor to see what is under the hood.

Unboxing Harbor

After seeing some good reasons to consider using Harbor as a container artifact repository around business, security, operational control, and extensibility, let's learn what Harbor is made up of. In this section, we will learn about the internal components and functions of Harbor. Being a container registry to serve the cloud-native community, Harbor itself is a cloud-native application comprised of multiple smaller microservices performing different activities. Let's understand how they work together by providing an architectural overview of Harbor.

Architecture overview

Harbor has several internal and external components. As shown in *Figure 6.2*, we can distribute these components into the following categories:

- **Consumers**: Consist of all clients and client interfaces
- **Fundamental Services**: Consist of all core functionalities that are part of the Harbor project and other key third-party projects that are essential components of the overall package
- **Data Access Layer**: Consists of all the different data stores
- **Identity Providers**: Consist of all external authentication provider extensions
- **Scan Providers**: Consist of all external image CVE scanner extensions
- **Replicated Registry Providers**: Consist of all external image replication extensions:

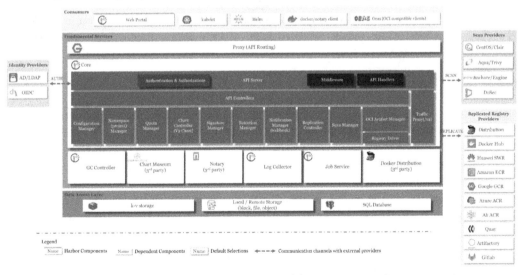

Figure 6.2 – Harbor 2.0 architecture (`https://github.com/goharbor/`
`harbor/wiki/Architecture-Overview-of-Harbor`)

Let's review some of the key components covered in *Figure 6.2*. You will see these components deployed in your Kubernetes environment when we install and configure Harbor later in this chapter.

Harbor Chart Museum

As we discussed previously, along with images, Harbor also supports storing Helm charts. To support this feature, Harbor internally uses **Chart Museum**, an open source project for the Helm chart repository. Visit `https://chartmuseum.com/` for more details. You can see this project deployed as a Kubernetes deployment resource named `my-harbor-chartmuseum` with a Kubernetes service running with the same name once you have a running instance of Harbor in your Kubernetes cluster.

Harbor Core

As described in *Figure 6.2*, Harbor Core is a collection of several modules that include key capabilities of Harbor being a container registry. These capabilities include concerns such as API management, authentication and authorization, interfacing glues, including pluggable image replication providers, image scanners, and image signature providers, and other foundational functionalities such as multitenancy capabilities, configuration management, artifact manager, and others. In our Kubernetes-based Harbor deployment, all the modules displayed in *Figure 6.2* under **Core** are deployed as a Kubernetes deployment resource named `my-harbor-core`, and this is exposed as a Kubernetes service resource with the same name.

Harbor job service

This is Harbor's asynchronous task execution engine that exposes the required REST APIs for other components to submit their job requests. One such example is a job to scan an image. You will see this microservice also getting deployed as its own Kubernetes deployment and a service named `my-harbor-jobservice`.

Harbor Notary

Notary (`https://github.com/notaryproject/notary`) is a third-party open source project under the CNCF umbrella. It is used to provide content trust establishment capabilities, which are achieved through an image signing procedure. As reviewed under the security-related reasons to use Harbor, such an image signing capability could be a great way to ensure that only verified images are deployed in a Kubernetes environment. It allows the image publisher to digitally sign an image using a private key authenticating the signer. Then, the consumers of that image can verify the publisher/signer of the image and take an informed decision to either trust or distrust the image based on the digital signature and the associated metadata. In secured and fully automated Kubernetes platforms, such operations of image signing and their verification are the steps of a CI/CD pipeline. Notary provides this functionality using its two main components – the server and the signer. The Notary server is responsible to store content metadata, ensuring the validity of the uploaded content, attaching the timestamps, and serving this content to the clients when requested. On the other side, the Notary signer is responsible for storing the private signing keys in a separate database from the Notary server database and performing the signing operations using these keys as and when requested by the Notary server. You will see these two components deployed as Kubernetes deployment resources named `my-harbor-notary-server` and `my-harbor-notary-signer`, along with their corresponding service resources, in a Kubernetes-based Harbor deployment that we will cover later in this chapter.

Harbor portal

As the name suggests, it is the **graphical user interface** (**GUI**) for Harbor. It provides sophisticated screens to perform all image registry and administrative configuration activities. In addition to this portal, Harbor also supports all operations using a REST API interface for automation requirements. You will see this component deployed as a Kubernetes deployment resource named `my-harbor-portal`, along with its corresponding service with the same name, later in this chapter.

Harbor registry

This is based on the open source project named Distribution (`https://github.com/distribution/distribution`), which wraps functionalities to pack, ship, store, and deliver content. It implements the standards defined by the OCI Distribution Specification. It is the core library used for image registry operations and used by many open source and commercial registries, including Docker Hub, GitLab Container Registry, and DigitalOcean Container Registry, including Harbor. You will see this component deployed as a Kubernetes deployment resource named `my-harbor-registry`, along with its exposed service with the same name, later in this chapter.

PostgreSQL database

This is the main database of Harbor and is used to store all required configurations and metadata used in Harbor. It stores all Harbor constructs, including but not limited to the data related to projects, users, policies, scanners, charts, and images. It is deployed as a stateful set on a Kubernetes cluster called `my-harbor-postgresql`, along with its service resource exposed with the same name.

Redis cache

This is also deployed as a stateful set on Kubernetes and it is called `my-harbor-redis-master`. It is used as a key-value store to cache the required metadata used by the job service.

Trivy Scanner

This is an open source project by Aqua Security (`https://github.com/aquasecurity/trivy`) and the default image CVE scanner that is deployed with Harbor 2.x. It can scan operating system layers and language-specific packages that are used in the image to find known vulnerability exposures present in those artifacts. Harbor uses such scanners to provide a comprehensive scanning capability. Such scanners can scan images and produce detailed reports, including CVE metadata. Such metadata includes a list of CVE numbers, vulnerability areas, severity levels, fixed versions if available, and other details. You will see this component getting deployed as `my-harbor-trivy` as a Kubernetes deployment post our installation.

> **What is my-harbor?**
>
> The prefix, `my-harbor`, that you have seen in the names of different components that will be deployed in your Kubernetes cluster is an arbitrary name given to the Helm chart instance of Harbor at the time of deployment. It can be replaced with any other name.

There are several other internal and external components described in *Figure 6.2* other than what we have covered here. The components we have covered are based on what is deployed on our Kubernetes cluster under Harbor's namespace. To learn more details about Harbor's architecture, visit this link: `https://github.com/goharbor/harbor/wiki/Architecture-Overview-of-Harbor`.

Now that we've learned about the different modules of Harbor, let's learn how to install and configure it on a Kubernetes cluster.

Getting started with Harbor

In this section, we will learn how to install and configure a Harbor registry instance on an existing Kubernetes cluster. But before we do that, we need to ensure that the following prerequisites are met.

Prerequisites

The following are the prerequisites for the Harbor installation instructions given in this section:

- Kubernetes cluster with version 1.10+

- Open internet connectivity from the Kubernetes cluster

- The operator machine should have the following tools:

  - `docker` CLI: `https://docs.docker.com/get-docker/`

  - `helm` CLI version 2.8.0+: `https://helm.sh/docs/intro/install/`

  - `kubectl` CLI version 1.10+: `https://kubernetes.io/docs/tasks/tools/`

- There should be a default **StorageClass** configured in your Kubernetes cluster that Harbor can use to create required storage volumes. By default, Harbor will need several **PersistentVolumeClaim** resources that are used by Redis cache, a PostgreSQL database, the registry storage, and more.

- The infrastructure running the Kubernetes cluster should be able to expose an externally accessible IP address upon the creation of a `LoadBalancer` type Kubernetes service, making it accessible outside the Kubernetes cluster. We have used **Google Kubernetes Engine** (**GKE**) in this chapter, which automatically creates a load balancer instance on GCP to front the `LoadBalancer` type service deployed in the GKE cluster.

- The operator machine should have a browser to access the Harbor GUI.

> **Additional learning**
>
> To keep the deployment of Harbor on Kubernetes simpler, we could also deploy it as a `NodePort` service and access it externally using the Kubernetes node IP address and the port associated with the Harbor service. However, we cannot access this deployment of Harbor from a Docker client to push and pull images using the node port. This is because the Docker client can only connect to a registry using port `443` (HTTPS) or port `80` (HTTP).
>
> Deploying a load balancer machine or a service instance for each `LoadBalancer` type service running on a Kubernetes cluster is not an efficient approach when several `LoadBalancer` type services are running on a Kubernetes cluster. Because, in this way, we may need several external load balancer instances for each externally facing service in the Kubernetes cluster. It is especially inefficient in a public cloud environment such as GKE, where such load balancer instances are charged separately. In a more sophisticated way, we can expose such externally facing services outside of a Kubernetes cluster using an **Ingress Controller** service running in the Kubernetes cluster. **Contour** (`https://projectcontour.io/`) is one such open source project under CNCF to be an Ingress Controller used for this reason that is supported by VMware and supplied with **Tanzu Kubernetes Grid**, which we will cover in the next chapter.
>
> To keep things simple for learning, we have used GKE to expose Harbor externally for this chapter. However, AWS Elastic Kubernetes Service and Azure Kubernetes Service can also provision the load balancers, similar to GKE. If your Kubernetes cluster is running on an infrastructure that cannot automatically expose a `LoadBalancer` service using an external endpoint, you can also do that manually. For that, you need to create a reverse-proxy server such as Nginx and deploy Harbor as a NodePort service rather than a `LoadBalancer` service using the `--set service.type=NodePort` option for the `helm install` command for Harbor deployment, which will be covered later in the installation steps.

Now, let's start installing Harbor.

Installing Harbor

While there are various ways to install and configure the Harbor repository, we will use a simple and easy-to-follow Bitnami-provided Helm chart approach to get a Harbor instance up and running in a Kubernetes cluster. It is required that all the steps in this section are performed using the same workstation. Let's get started:

1. Add Bitnami's Helm repository to your workstation:

   ```
   $ helm repo add bitnami https://charts.bitnami.com/
   bitnami
   "bitnami" has been added to your repositories
   ```

2. Create a namespace on the cluster to deploy all Harbor components within it:

   ```
   $ kubectl create namespace harbor
   ```

3. Install the Helm chart to get Harbor components deployed in the `harbor` namespace. It should deploy all Harbor components to expose a `LoadBalancer` type Kubernetes service to expose the portal:

    ```
    $ helm install my-harbor bitnami/harbor -n harbor
    ```

    Within just a few seconds, you will see the following output if the command was successfully executed. Here, `my-harbor` is just a name given to this Helm deployment that can be used to upgrade or delete the installation with that name reference:

    ```
    NAME: my-harbor
    LAST DEPLOYED: Mon Mar 28 22:36:23 2022
    NAMESPACE: harbor
    STATUS: deployed
    REVISION: 1
    TEST SUITE: None
    NOTES:
    CHART NAME: harbor
    CHART VERSION: 12.2.4
    APP VERSION: 2.4.2


    ** Please be patient while the chart is being deployed **


    1. Get the Harbor URL:

        echo "Harbor URL: https://127.0.0.1:8443/"
        kubectl port-forward --namespace harbor svc/my-harbor
    8443:443


    2. Login with the following credentials to see your
    Harbor application

        echo Username: "admin"
        echo Password: $(kubectl get secret --namespace harbor
    my-harbor-core-envvars -o jsonpath="{.data.HARBOR_ADMIN_
    PASSWORD}" | base64 --decode)
    ```

    The preceding code snippet shows the output of the successful execution of the `helm install` command.

4. Check the status of all the pods deployed in the `harbor` namespace to ensure everything is running fine:

```
$ kubectl get pods -n harbor
```

You should see all the pods running and all the containers in a ready state, as shown in the following code snippet:

```
NAME                                         READY   STATU
S    RESTARTS        AGE
my-harbor-chartmuseum-64bdb5df6f-js-
flz      1/1      Running   0                5m38s
my-harbor-core-54fb5d55c9-
jjw8p            1/1      Running   1 (3m23s ago)   5m38s
my-harbor-jobser-
vice-66894459d4-rtv4n       1/1      Running   2 (3m28s
ago)    5m38s
my-harbor-nginx-54f6b75d9b-
9dtxt            1/1      Running   0                5m38s
my-harbor-notary-serv-
er-5fc48d989c-b5s96   1/1      Running   0                5m38s
my-harbor-notary-signer-5bff-
c645b-26r4c      1/1      Running   1 (3m38s ago)   5m38s
my-harbor-portal-697d84949-
k7hrv            1/1      Running   0                5m38s
my-harbor-post-
gresql-0             1/1      Running   0                5m38s
my-harbor-redis-mas-
ter-0                1/1      Running   0                5m38s
my-harbor-regtry-6f55746685-
rr4ts        2/2      Running   0        5m38s
my-har-
bor-trivy-0          1/1      Running   0                5m38s
```

5. Retrieve the admin user password generated by the installation. Note down this password as we will use it later to access the Harbor GUI:

```
$ echo "admin password: $(kubectl get secret --namespace
harbor my-harbor-core-envvars -o jsonpath="{.data.HARBOR_
ADMIN_PASSWORD}" | base64 --decode)"
```

6. Retrieve the external IP address on which Harbor service is exposed by running the following command:

```
$ echo "Harbor Service IP: $(kubectl get service my-har-
bor -n harbor -o jsonpath='{.status.loadBalancer.
ingress[0].ip}')"
```

7. Access the Harbor portal from a browser using `https://<external-ip>/`. You may need to ignore the browser security prompts as we are not using a certificate signed by a valid certificate authority for the portal. You should see the following screen:

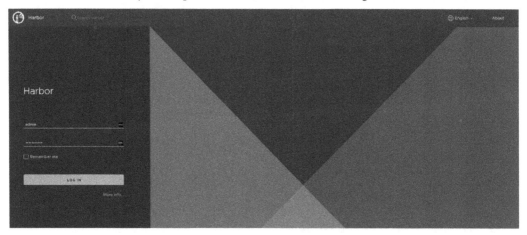

Figure 6.3 – Harbor login page

8. Enter the username as `admin` and the password that you retrieved in *step 5* previously. You should be able to log in successfully and see the following home page of the Harbor GUI:

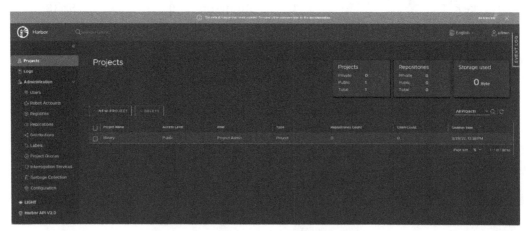

Figure 6.4 – Harbor landing page

If you can log in using the aforementioned credentials and see the previous screen, you are done with the required setup to have a Harbor instance on your Kubernetes cluster running. As the next step, we will perform a small smoke test by pushing an image to this registry to validate our setup.

Validating the setup

Perform the following steps to validate the installation:

1. Retrieve the CA certificate used by Harbor by following these steps. We need to add this certificate in the trust store used by the Docker client to connect to the Harbor deployment:

    I.   Go into the **library** project by clicking on the highlighted link:

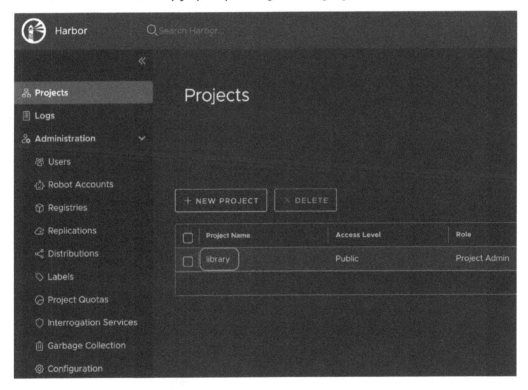

Figure 6.5 – Clicking the library project

    II.   Download the certificate into your workstation using the **REGISTRY CERTIFICATE** link shown in the following screenshot:

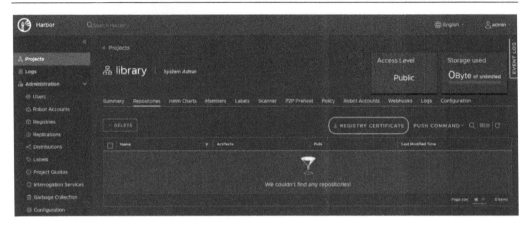

Figure 6.6 – Downloading the registry certificate

2. Import the Harbor portal certificate in the trust store used by the Docker client running in your workstation as explained here, depending on your operating system: `https://docs.docker.com/registry/insecure/#use-self-signed-certificates`.

    For macOS, run the following command:

    ```
    $ sudo security add-trusted-cert -d -r trustRoot -k /
    Library/Keychains/System.keychain <downloaded-harbor-
    certificate>
    ```

    Here, replace `<downloaded-harbor-certificate>` with the certificate path that was downloaded in *step 1*.

3. Restart the Docker daemon in your workstation if it is running; otherwise, start it to make the certificate visible to the Docker client.

4. Create a local DNS entry in the `/etc/hosts` file to link the default domain name, `core.harbor.domain`, with the external load balancer IP address that was used to access the portal in the previous steps.

5. Log in to the Harbor registry using the `docker` CLI to enable push/pull operations:

    ```
    $ docker login -u admin https://core.harbor.domain
    Password: <as retrieved from the step #5 of installation>
    Login Succeeded
    ```

6. Push an image into your newly setup Harbor registry using the Docker client:

    I.   Download the `busybox:latest` image using the `docker` CLI. The following command will download the image from the Docker Hub repository into your local workstation:

    ```
    $ docker pull busybox:latest
    ```

II.   Verify the presence of the `busybox:latest` image in your local image repository by running the following command:

```
$ docker images | grep busybox
```

III. You should see a record indicating the `busybox` image with a `latest` tag as a result of the previous command.

IV.   Tag the `busybox:latest` image to prepare it to push to our Harbor registry instance:

```
$ docker tag busybox:latest core.harbor.domain/library/
busybox:latest
```

V.   Push the newly tagged image to your Harbor instance:

```
$ docker push core.harbor.domain/library/busybox:latest
```

7. Upon this successful push operation, you should be able to see the image listed in your **library** project in the portal, as shown in the following screenshot:

Figure 6.7 – Verifying the new image presence

If you see the `busybox` image in the previous screen, your Harbor installation is complete. You may also choose to perform a pull test either from the same workstation by removing the existing image from the local repository, or a different workstation that has a Docker client. If you prefer to use a different workstation, you may need to configure the Harbor certificate there and authenticate against the Harbor repository using the `docker` CLI.

In the next section, we will cover some of the crucial day-2 operations for Harbor using this installation.

Common day-2 operations with Harbor

Now that we have a working setup of Harbor, let's look into some important day-2 operations that we may need to perform on it. We will cover the following activities in this section:

- Creating and configuring a project in Harbor, which is the multi-tenancy enabling construct on Harbor that allows different teams to have separate configurations
- Configuring automated image scanning and working with the scan results
- Preventing insecure images from being used to deploy containers using them
- Configuring image replication to allow selective access to the external images
- Performing a cleanup of unused image tags to free up the storage quota of a project

As you can see here, we have lots of ground to cover. So, let's get started.

Configuring a project in Harbor

Let's create a new project called **project-1** and configure it using the admin user we used previously to verify the installation:

1. Click on the **NEW PROJECT** button on the home screen of the Harbor portal after logging in as **admin**:

Figure 6.8 – Creating a new project

2. Configure the project's name and quota, as shown in the following screenshot, and click **OK**. We will keep this project private, which means that you need to get authenticated to pull images. We will also not configure this project to be used as a pull-through cache for an external registry such as Docker Hub:

Figure 6.9 – Entering New Project details

3. You should be able to see a new project listed on the screen, as shown in the following screenshot:

Figure 6.10 – Verifying the new project's presence

Now that we have this project created, the users of this project may push their images with the `core.harbor.domain/project-1/` prefix for the images. The project will not accept new images after it reaches its 2 GB storage quota, which we configured while creating it in *step 2*. Now, let's learn about some of the important project-level configurations.

Configuring image scanning for a project in Harbor

Scanning images for the presence of CVEs is an important security feature of Harbor. Now, let's configure **project-1** to enable automated CVE scanning as soon as an image is pushed in this project using the default scanner, **Trivy**:

1. Go to the **project-1** detail page by clicking on the highlighted link:

Figure 6.11 – Selecting the newly created project

2. Open the **Configuration** tab and select the highlighted option to scan every image upon push. Finally, click **SAVE**:

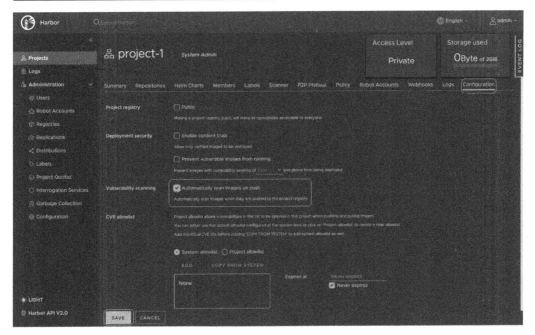

Figure 6.12 – Enabling image auto-scanning for the project

Now, let's verify whether this configuration works by pushing the busybox:latest image that we pulled from Docker Hub previously. This should be present in your local workstation's Docker repository.

3. Prepare the busybox:latest image to be pushed into the **project-1** repository of Harbor by applying the appropriate tag:

```
$ docker tag busybox:latest core.harbor.domain/project-1/
busybox:latest
```

4. Push the newly tagged busybox:latest image to the **project-1** repository:

```
$ docker push core.harbor.domain/project-1/busybox:latest
The push refers to repository [core.harbor.domain/
project-1/busybox]

797ac4999b67: Layer already exists
latest: digest: sha256:14d4f50961544fdb-
669075c442509f194bdc4c0e344bde06e35dbd55af842a38 size:
527
```

As you can see in the result of the previous command, Harbor used the same layer of the image that we had pushed under the **library** project during the verification process we performed earlier.

erify the image presence under **project-1** and click on the **project/busybox** link that is highlighted in the following screenshot:

Figure 6.13 – Clicking the image repository

5. As shown in the following screenshot, the scanning of the busybox:latest image has already been completed with no vulnerabilities found in it. This scanning was triggered automatically upon pushing the new image into the repository:

Figure 6.14 – Verifying the scanning results

Now, let's create a policy based on such scan results to prevent pulling an image that has CVEs of more than medium severity to prevent running a container with such vulnerabilities.

Preventing insecure images from being used in Harbor

To stop insecure images from being pulled, navigate to the project landing page of **project-1**. You may see an option named **Prevent vulnerable images from running** under the **Configuration** tab, as shown in the following screenshot. Check this option, select **High** from the dropdown menu, and save the configuration changes:

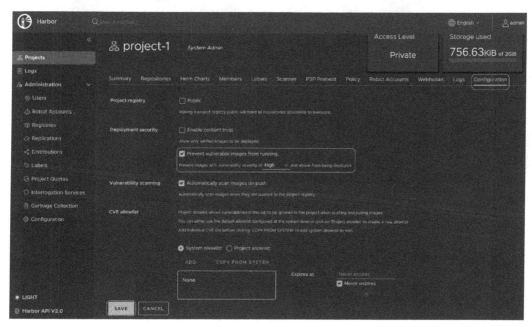

Figure 6.15 – Preventing image pulling with high and critical CVEs

Now, let's test this configuration change. For that, we need to push an insecure image into **project-1**. You may follow these steps to perform this test:

1. Pull the `nginx:1.9.5` image from Docker Hub, which is very old and full of CVEs:

   ```
   $ docker pull nginx:1.9.5
   ```

2. Tag `nginx:1.9.5` for the Harbor **project-1** repository:

   ```
   $ docker tag nginx:1.9.5 core.harbor.domain/project-1/
   nginx:1.9.5
   ```

3. Push `nginx:1.9.5` to the Harbor **project-1** repository:

    ```
    $ docker push core.harbor.domain/project-1/nginx:1.9.5
    ```

4. Verify that the image on Harbor under **project-1** has been scanned and showing CVEs, as shown in the following screenshot:

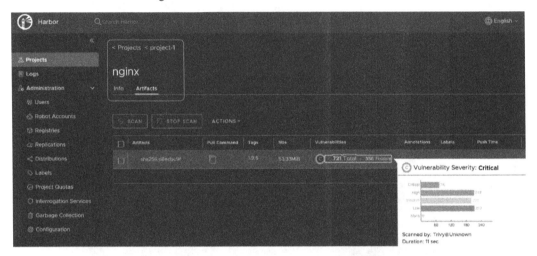

Figure 6.16 – Verifying the CVE scan results

5. Delete the `nginx:1.9.5` image from the local Docker repository so that we can attempt to pull it from our Harbor **project-1** repository:

    ```
    $ docker image rm core.harbor.domain/project-1/
    nginx:1.9.5 nginx:1.9.5
    ```

6. Try to pull `nginx:1.9.5` from the Harbor **project-1** repository:

    ```
    $ docker pull core.harbor.domain/project-1/nginx:1.9.5


    Error response from daemon: unknown: current image with
    721 vulnerabilities cannot be pulled due to configured
    policy in 'Prevent images with vulnerability severity of
    "High" or higher from running.' To continue with pull,
    please contact your project administrator to exempt
    matched vulnerabilities through configuring the CVE
    allowlist.
    ```

As you can see, Harbor denied sending the image because of the CVEs present in the image above the configured tolerance threshold.

> **Important note**
>
> We took this testing approach to keep it simple. But alternatively, you may also attempt to create a pod using this image on a Kubernetes cluster to mimic a practical scenario. The result would be the same and you may not create a pod using the `core.harbor.domain/project-1/nginx:1.9.5` image. However, to test creating a pod using this Harbor setup, you may need to add the DNS entries to all your Kubernetes cluster nodes' `/etc/hosts` file. Alternatively, you might need to create a more production-like Harbor setup with a proper domain name that can be resolved using an external DNS record from within the Kubernetes cluster.

This concludes our configuration and its test to prevent insecure container images from being pulled from a Harbor repository. In the next section, we will learn how to configure a remote repository sync to allow internal developers to pull the required externally available allowed images via Harbor.

Replicating images in Harbor

In this section, we will learn how to configure image replication in Harbor. There are two types of replications in Harbor, as described here, along with their practical applications:

1. **Push-based**: To configure a rule to push certain images from Harbor to another repository, which could be another remote Harbor deployment or even a public repository. This type of replication is very useful to implement a hub-and-spoke type of deployment where we need to make certain images available in the Kubernetes clusters running at edge locations. Having the required images on the same Kubernetes clusters available at the edge location could be a great help to reduce network latency and dependency (hence application availability) when the image pulls are required to deploy containers on the edge Kubernetes clusters. This scenario is depicted in *Figure 6.1*.

2. **Pull-based**: To configure a rule to pull certain images from a remote public or private repository. Such a replication policy allows access to developers of certain allowed container images from an external repository such as Docker Hub or GCR. This feature of Harbor not only allows freedom for developers to use approved externally hosted images but also allows operators to prevent the wild-wild-west situation where anyone may pull any image from an external repository. *Figure 6.1* shows how the central Harbor repository pulls required images from Docker Hub and GCR using this feature.

Now that we understand the types and their uses, let's see how we can configure these replication rules in Harbor. Here, we will configure a pull-based policy to allow developers to access MySQL images from Docker Hub. We will configure a remote repository location and the replication rule in Harbor, followed by quickly verifying that these configurations are working as expected:

1. Add Docker Hub as an external registry endpoint:

    I.   Click on the **Registries** option under the **Administration** menu and click on the **NEW ENDPOINT** button, as shown in the following screenshot:

Figure 6.17 – Adding a new external registry endpoint

    II.   Select **Docker Hub** from the dropdown, enter a name and description, and click on the **TEST CONNECTION** button. You may leave the authentication details empty for this test. However, for a production-grade deployment, you should supply these credentials to prevent Docker Hub from applying an image pull rate limit. Docker Hub throttles unauthenticated pull requests with lower rate limits:

Figure 6.18 – Adding Docker Hub details and testing the connection

III. Click the **OK** button to save and exit.

IV. Verify the newly created entry under the **Registries** page, as shown in the following screenshot:

Figure 6.19 – Verifying the presence of the Docker Hub endpoint

2. Create a replication rule that allows us to pull MySQL images from Docker Hub using the registry endpoint we just created:

I.   Click on the **Replications** option under the **Administration** menu and click on the **NEW REPLICATION RULE** button, as shown in the following screenshot:

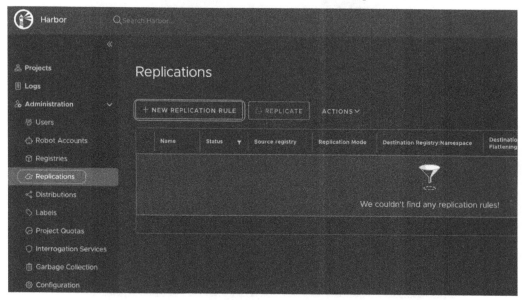

Figure 6.20 – Creating a new replication rule

II.   Enter the name and the description of the replication rule, select the **Pull-based** replication option, select the Docker Hub registry endpoint from the dropdown that we created in *step 1*, provide image filter criteria, as shown in the following screenshot, and click the **SAVE** button:

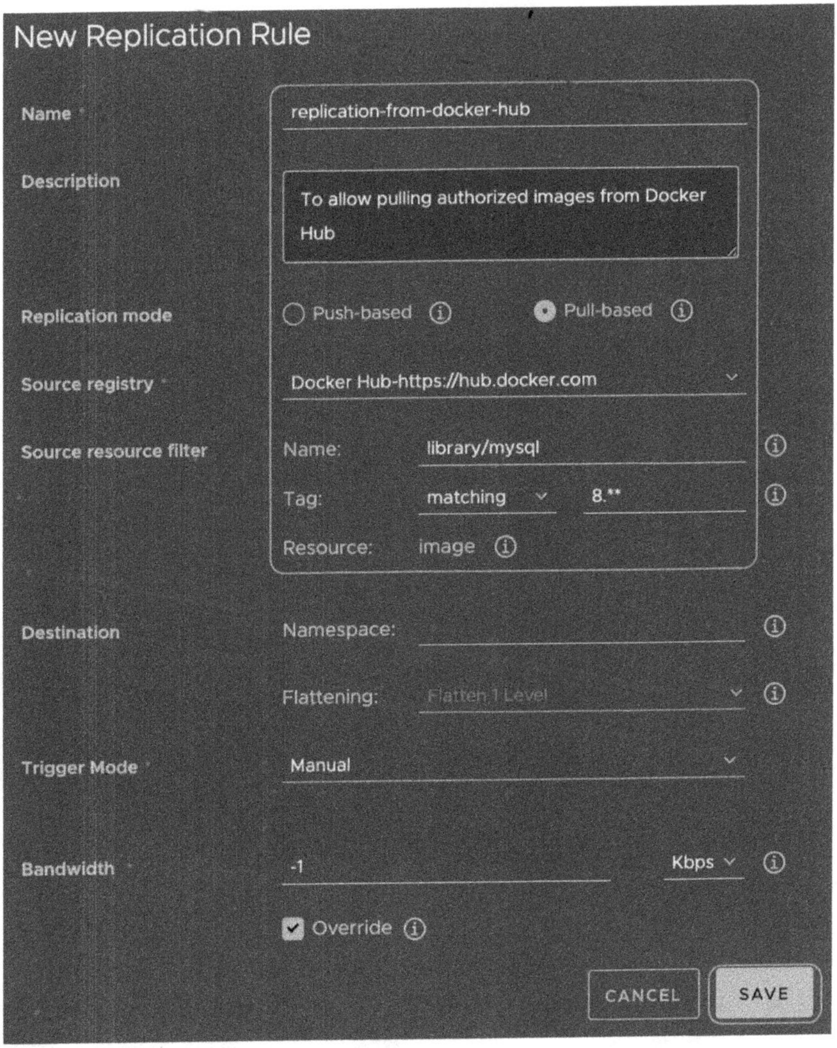

Figure 6.21 – Submitting the replication rule's details

We will leave the other options as-is. These other options include the image destination details to specify if we want a different image folder name and the directory structure, which is different from the source. It also has the option to select when we want to trigger pulling the images matching the filter criteria. The default value of the same is to pull manually when required but we can also create a schedule-based pull. Then, we have the option to restrict the bandwidth requirements to prevent the network from being overwhelmed with a flood of pull requests being executed for large filter criteria. And finally, there is the option to enable/disable the images from being overwritten when there is an image with a different SHA but the same name and tag available on the source registry endpoint.

> **Customized filter patterns**
>
> In the previous configuration, we used very basic filter criteria to pick all MySQL images that have tags starting with `8.` characters. However, you may apply complex patterns here to allow/disallow images based on your requirements. You may learn more about such patterns here: `https://goharbor.io/docs/2.4.0/administration/configuring-replication/create-replication-rules/`.

III. Trigger the replication manually by selecting the newly created rule and clicking on the **REPLICATE** button:

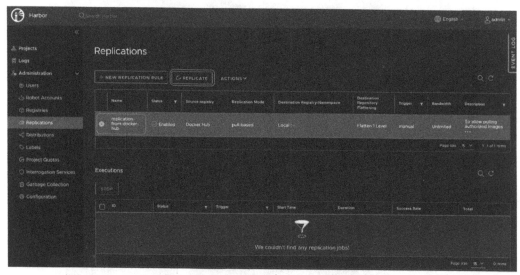

Figure 6.22 – Triggering image replication

IV. Confirm the replication by pressing the **REPLICATE** button:

Figure 6.23 – Confirming image replication

V.   Verify whether the replication execution was successful from the **Executions** section on the **Replications** page, as shown in the following screenshot. Depending on the network connection, it may take a few minutes before all the images are successfully replicated:

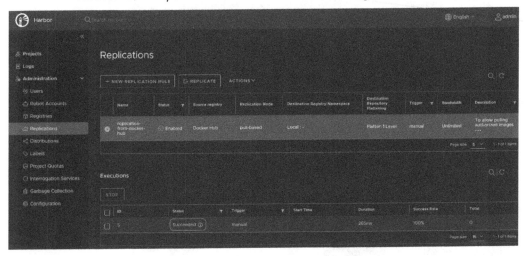

Figure 6.24 – Verifying image replication execution

VI.   Upon successful completion of this replication, go to the **Projects** screen and click on the **library** project:

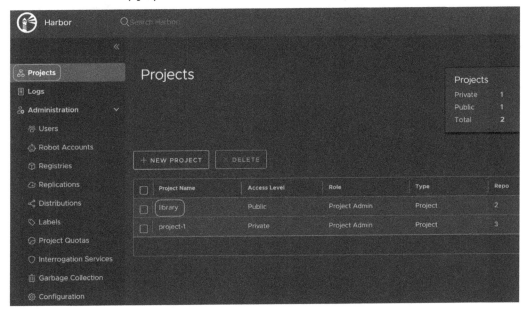

Figure 6.25 – Opening the library project

VII. You will see a new namespace under the **Repositories** tab named **library/mysql**. Click on that link:

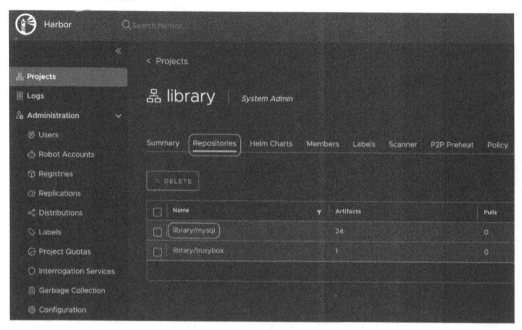

Figure 6.26 – Verifying the newly replicated repository

VIII. You may see several images listed under **library/mysql** as a result of the replication operation:

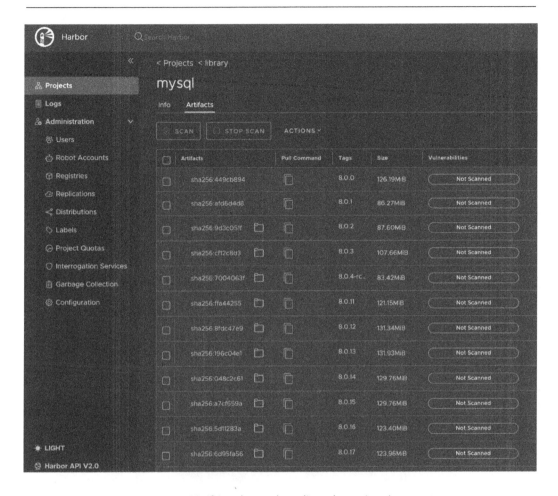

Figure 6.27 – Verifying the newly replicated repository's content

IX.   Pull one of these images into the local workstation's Docker repository to verify it is working as expected:

```
$ docker pull core.harbor.domain/library/mysql:8.0.27
```

This command should be able to pull the image successfully from our Harbor repository.

This concludes our replication configuration step. Now, the developers may pull a required MySQL image directly from the Harbor repository rather than getting them from Docker Hub.

As the next day-2 activity in this list, we will learn how to configure a rule-based tag retention policy to clean up stale images and free up the storage.

Configuring rule-based tag retention policies in Harbor

As you saw in the previous configuration, we replicated over 25 different images for MySQL. In production-grade implementations, we often encounter situations where there are several stale images not being used but occupying the project's allocated space quota. In our previous MySQL replication, we may see only a few image tags that are used, and we can remove the rest. For that, we will learn how to configure such automated tag-based retention policies to clean up old and useless content. Let's get started:

1. Go to the **Projects** screen and click on the **library** project:

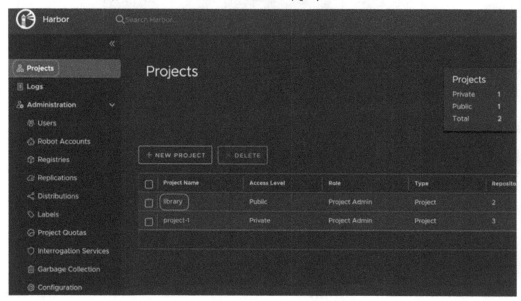

Figure 6.28 – Opening the library project

2. Under the **library** project, go to the **Policy** tab and click on the **ADD RULE** button:

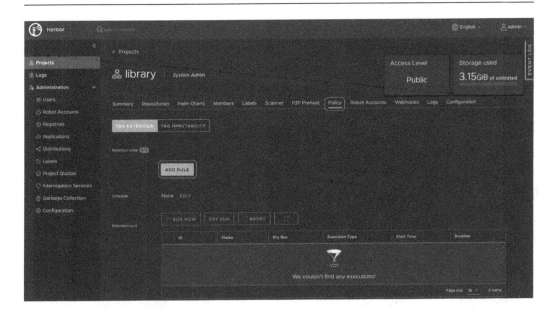

Figure 6.29 – Adding a new image retention rule

3. Enter `mysql` as the matching repository, select **retain the most recently pulled # artifact** from the dropdown, as shown in the following screenshot, enter `1` as the count to retain, and press the **ADD** button:

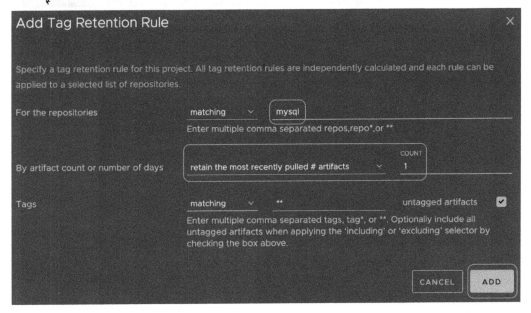

Figure 6.30 – Entering details about the image retention rule

This policy will remove all MySQL images except for `mysql:8.0.27` because we pulled that one in the verification step earlier.

4. As you can see, the new tag retention policy is in place. While we can also schedule this activity at a regular frequency, for now, we will run it manually using the **RUN NOW** button:

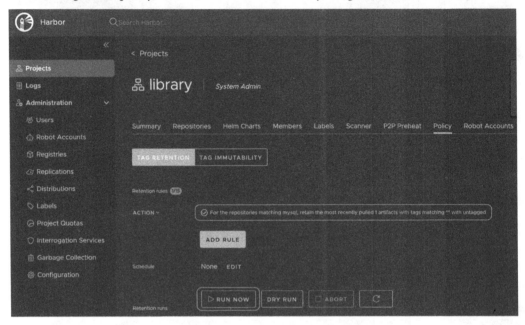

Figure 6.31 – Verifying the creation of the image retention rule

5. Move on by pressing the **RUN** button by accepting the warning of a mass deletion operation:

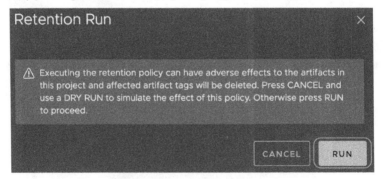

Figure 6.32 – Executing the image retention rule

6. Verify whether the clean-up activity was completed successfully by inspecting the activity log record, which indicates there was only 1 tag retained out of 24 total. This was the expected outcome:

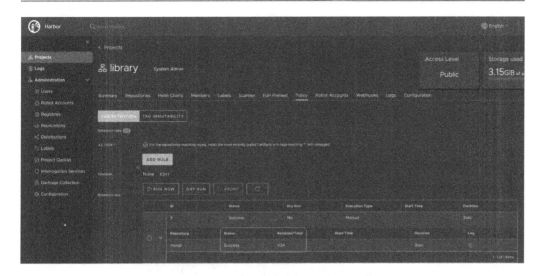

Figure 6.33 – Verifying the image retention rule's execution

7. Click on the **Repositories** tab to list the existing repositories. On that screen, you should be able to see only 1 tag available for **mysql** instead of the 24 from earlier:

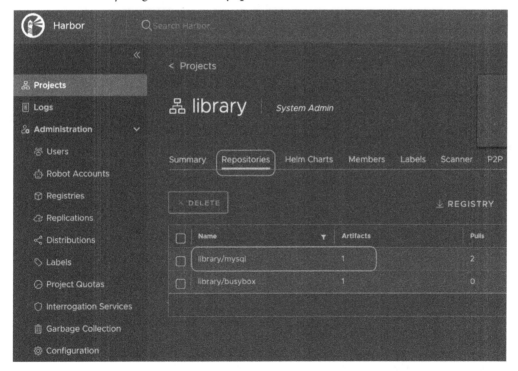

Figure 6.34 – Verifying the remaining image count post-retention rule's execution

8. Navigate to the `mysql` repository to verify that the tags other than `8.0.27` have been deleted:

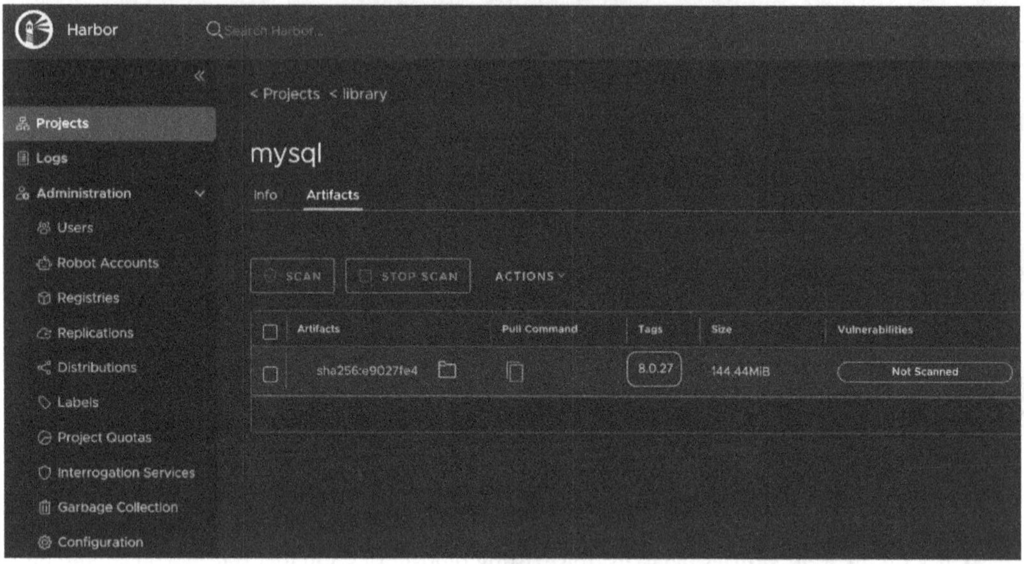

Figure 6.35 – Verifying the remaining image's post-retention rule's execution

This concludes our tag retention policy configuration. Like tag retention, we may also configure tag immutability policies to configure certain critical image tags from being overwritten with a new version of the image. Without such a policy in place, you may push new images with changes but with the same tag value. Hence, an application that is tested against one tagged version of an image may not be able to fully ensure that the content of the same tag would be the same in the next pull of the image. This could potentially break applications from working in case of any unexpected changes in the newer version of the same tagged images are pushed. Ideally, this should not be the case as it is a poor development practice. But there should be some controls in place to prevent it from happening. Hence, Harbor helps in this case by allowing you to create policies where the Harbor users may not push images with the same tag with different content determined by the SHA-256 algorithm.

> **Important information**
>
> To safeguard our containerized applications from failing because of changed image content for the same tag value, it is always a good practice to pull images using their SHA values rather than their tags. Pulling an image using its SHA value (also known as the digest) ensures you always get the same image with the same content and there is no fear of it getting accidentally overwritten with the same tag value. For example, the image content pulled with the `docker pull ubuntu:20.04` command can be theoretically different for multiple executions, but the image content will always be the same when it is pulled with its digest using the `docker pull ubuntu@sha256:82becede498899ec668628e7cb0ad87b6e1c371cb8a1e-597d83a47fac21d6af3` command.

There are several more administrative and user day-2 activities we may need to perform on Harbor, including user account management, robot accounts for automation, image signing, webhook-based automation triggers, and many more. In this chapter, we covered some of the most common activities that we usually perform on a container registry such as Harbor. While we can add more details for other operations, the goal of this book is not to be a Harbor guide alone. But if you want, you may find more details in the official documentation for Harbor at `https://goharbor.io/docs/2.4.0/`.

Summary

Let's review what we have covered in this chapter. First, we discussed several benefits and the use cases of Harbor while explaining the *Why* behind using it. We looked at the different security-related benefits of Harbor, including image scanning, robust RBAC capabilities, and the ability to restrict public repository access requirements using image replications. For the operational control aspect, we discussed the benefits, such as on-premises and air-gapped deployment, a popular open source project under CNCF, comprehensive multi-tenancy, and administrative configurations. For the extensibility aspect, we saw how Harbor can be used with its replication feature of extending image library contents. Harbor's pluggable model for vulnerability scanners and authentication providers was also discussed in this category.

After that, we covered details of Harbor's architecture and learned about the different components that make up Harbor in detail. Following this, we learned how to quickly get started with Harbor using the Bitnami Helm chart and verified the installation. Finally, we walked through some of the important day-2 operations around Harbor, including creating a project, performing image scanning, preventing risky images from being pulled, image replication from Docker Hub, and cleaning stale images to free up the storage quota for a project.

In the next chapter, we will learn how to run these images using Tanzu Kubernetes Grid, a multi-cloud Kubernetes offering from VMware.

7

Orchestrating Containers across Clouds with Tanzu Kubernetes Grid

In the previous chapter, we learned about Harbor, a container registry, that is covered as a part of the Tanzu product bundle. After learning about hosting our container images with Harbor, let's learn how to deploy them on Kubernetes in this chapter with Tanzu Kubernetes Grid, a multi-cloud and enterprise-ready Kubernetes distribution of Tanzu. Kubernetes has become widely accepted and the default norm in the industry to run containers in the past few years. As per a recent survey by the **Cloud Native Computing Foundation** (**CNCF**), 96% of the responding organizations were either evaluating or already using Kubernetes (source: `https://www.cncf.io/wp-content/uploads/2022/02/CNCF-Annual-Survey-2021.pdf`)! Additionally, a large sum of them used Kubernetes on different cloud platforms for reasons such as risk mitigation, avoiding vendor lock-ins, and efficient operational expenditure. But operating large Kubernetes platforms on one or more clouds is a non-trivial effort for various reasons. Each cloud platform has APIs and distinct ways to manage Kubernetes services. The complexity increases when the organizations also need to run Kubernetes services on-premises. For such on-premises Kubernetes deployments, many enterprises build their own platforms with custom automation. While building such custom platforms using open source tools brings initial cost savings, maintaining them for a long time is very difficult when the original people who built the platforms move out of the organizations. Rather, the organizations could better utilize their talents to build more revenue-generating custom applications for their businesses by using an enterprise-grade product such as **Tanzu Kubernetes Grid** (**TKG**) for such below-value-line concerns.

TKG is VMware's Kubernetes distribution that comes with all the bells and whistles to deploy enterprise-grade container platforms on vSphere-based on-premises data centers, **Amazon Web Services** (**AWS**), and Azure public cloud infrastructure. In this chapter, we will learn about this product in detail and will cover the following topics:

- **Why Tanzu Kubernetes Grid?**: A walkthrough of the features and capabilities of TKG

- **Unboxing Tanzu Kubernetes Grid**: A detailed overview of the components and the concepts of TKG

- **Getting started with Tanzu Kubernetes Grid**: Learn how to install and configure TKG

- **Common day-2 operations with Tanzu Kubernetes Grid**: Learn how to perform various cluster life cycle activities with TKG

Let's get started by learning about the background of Tanzu Kubernetes Grid.

Why Tanzu Kubernetes Grid?

In a nutshell, TKG is an enterprise-supported flavor of the open source Kubernetes platform. Like many other distributions, TKG uses the upstream Kubernetes distributions without modifying the open source code base. However, there are a few good reasons why an enterprise would like to use TKG over the open source community – *vanilla* – distribution of Kubernetes. We'll explore those reasons in this section.

Multi-cloud application deployments

As per a survey done by Gartner (`https://www.gartner.com/smarterwithgartner/why-organizations-choose-a-multicloud-strategy`), over 81% of respondents said that they have a multi-cloud deployment or strategy. Enterprises have multi-cloud strategies for reasons such as avoiding vendor lock-ins and using the best services offered by the respective cloud provider. Additionally, the enterprises have applications that may not be deployed on a public cloud platform and hence deployed on-premises. Kubernetes offers a great option to allow application deployments in multi-cloud and hybrid cloud platforms. Once an application is ready to run in Kubernetes wrapped in a container, it can be deployed on any upstream conformant Kubernetes platform deployed in any cloud or data center. Kubernetes made the multi-cloud deployment of applications almost trivial. Kubernetes manages containerized applications well, but at its core, it does not know how to manage the infrastructure on which it is deployed. That makes the platform and infrastructure management an external concern. All major cloud providers offer Kubernetes as a service in addition to the option of just using their infrastructure to deploy a self-managed Kubernetes platform. However, every platform has its own interfaces and hence a different user experience. Additionally, every platform has its own flavors of operating systems, networks, storage, and compute-level differences. Managing large Kubernetes platforms itself is a hard problem to solve, which is amplified even more when we need to manage a multi-cloud environment. While learning the details of one cloud platform is not complex enough, now, the infrastructure and the platform teams need to learn the same for more than one cloud platform if they aim to deploy Kubernetes in a multi-cloud fashion. Because of these reasons and the complexities involved, several enterprises avoid going multi-cloud despite the involved risks in using just one platform.

TKG addresses these challenges by providing a uniform experience to operate Kubernetes on vSphere (for on-premises deployment), AWS, and Azure cloud platforms. One of the core components of TKG is the Cluster API (`https://cluster-api.sigs.k8s.io/`), a CNCF open source project that provides an abstraction layer on top of the infrastructure. The Cluster API deploys and manages Kubernetes nodes with the required storage and network configuration for the selected cloud platform. Additionally, TKG exposes its interfaces using the `tanzu` CLI by providing the same user interface to perform different Kubernetes cluster life cycle operations, irrespective of the infrastructure provider. TKG also bundles VMware-supported operating system layers for the three supported cloud providers. With all these characteristics, TKG provides an easy-to-consume multi-cloud Kubernetes platform.

Open source alignment

Another of the main benefits of using TKG is its usage of open source tools for different use cases. As its core component, TKG uses the upstream open source Kubernetes distribution maintained by the CNCF community. TKG does not fork the source code of any of its open source components to add its own flavor and customization. This characteristic provides TKG consumers with all the benefits of using open source software, including avoidance of vendor lock-ins via portability, and an avenue to enrich the functionalities to their needs via open source contributions.

Considering these benefits, all enterprises have wanted to use open source solutions as much as possible in the past few years. However, the cloud-native ecosystem is very crowded with open source products that solve similar problems and have different levels of maturity to be used in production setups. *Figure 7.1* is a screenshot taken from the CNCF website (`https://landscape.cncf.io/`) showing an extremely crowded space with tools solving different problems running containerized applications. Picking the right tool for the right problem with acceptable maturity, a vibrant community, and enterprise-level support is a challenging task. And the landscape is constantly changing by adding new solutions to the list very frequently. Adding more to the complexity, the compatibility of one tool working with another is not always the case. So, there are three possible solutions to this issue:

1. Enterprises would build their custom container platforms by carefully evaluating their options in this diverse and crowded landscape by running long proof-of-concept projects to narrow down their choices. While this would provide all the control and possible short-term cost benefits, there are some considerable drawbacks to this approach. One of them is that enterprises may need to spend a lot of productive time with their engineers to build a custom container platform. Finding right talents from the market and building required skills internally are hard and expensive challenges. Additionally, such skilled people would be in high demand in the market and may not stay with the enterprise for a long time to support what they have built internally. And then, such custom-built solutions quickly become an expensive liability for the organization.

2. Enterprises would choose all the solutions provided by a cloud provider, where the enterprises no longer need to spend time selecting the tools and worry about the support and compatibility of those tools. This way, an enterprise can build a robust and production-grade container platform using all the services provided by a single hyper-scaler such as AWS, Azure, or Google Cloud Platform. However, this approach could result in potential vendor lock-ins. The *divorce* would be very painful in the long term and it would be an expensive *marriage* with one cloud provider:

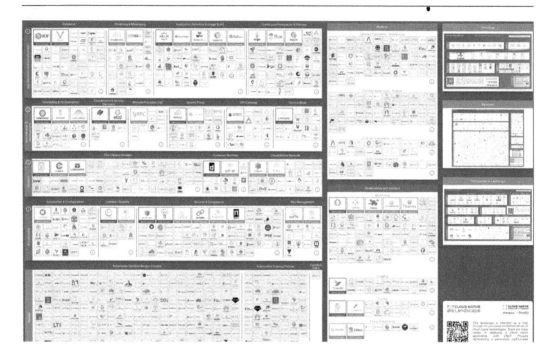

Figure 7.1 – CNCF landscape

3. Enterprises would pick a multi-cloud platform such as Tanzu that would abstract an infrastructure provider, resulting in the portability of workloads deployed on the platform. Like the first two approaches, this one also has a drawback. Using such multi-cloud platforms might result in a vendor lock-in situation for the vendor of the multi-cloud platform provider itself. There is no foolproof solution to avoid vendor lock-ins when we use any third-party product. However, this risk could be somewhat mitigated if the third-party solution is fully backed by open source components, which is exactly the case with TKG.

TKG is a bundle of many different popular open source projects, including Kubernetes as a container runtime, the Cluster API for Kubernetes cluster life cycle management, Antrea for container overlay networking, Ubuntu and Photon as the node operating systems, Fluent Bit for logging, Prometheus and Grafana for monitoring, and many others. We will cover them all in the next section. VMware supplies signed binaries for all these open source tools that work together for a given release of TKG. This way, TKG helps enterprises avoid going through the painful process of tool selection, makes them work together, gets them supported, and avoids vendor lock-ins to a good extent.

With this, we have answered the question *Why Tanzu Kubernetes Grid?* Now, let's understand what is included in the TKG bundle and its core concepts.

Unboxing Tanzu Kubernetes Grid

In this section, we will review all the building blocks of TKG, including its interface and core and extension components. After that, we will understand the core concepts of this platform to understand how it works. We have a long way to go, so let's start.

Building blocks of Tanzu Kubernetes Grid

As mentioned in the previous section, TKG is a collection of many open source tools that solve different problems that, together, make an enterprise-grade Kubernetes platform. We can distribute these components into three categories – interface, core, and extensions – as shown in *Figure 7.2*:

Figure 7.2 – Tanzu Kubernetes Grid bundle

Let's review all these components to learn about their roles in the TKG bundle.

Interface components of Tanzu Kubernetes Grid

As the name suggests, these components include TKG's interfaces with users and infrastructure providers, including vSphere, AWS, and Azure. The following section specifies these tools in more detail.

Tanzu command-line interface (CLI)

Like a few other products in the Tanzu product portfolio, TKG also uses the `tanzu` CLI as its primary user interface. The `tanzu` CLI has a plugin structure that allows different Tanzu products to use the same interface. TKG uses the `tanzu` CLI for all cluster operations, such as viewing, creating, scaling, deleting, and upgrading TKG clusters. We will use this CLI in the next section when we set up our TKG foundation. The `tanzu` CLI is a part of the broader open source project named Tanzu Framework (`https://github.com/vmware-tanzu/tanzu-framework`). You can learn more about this CLI here: `https://github.com/vmware-tanzu/tanzu-framework/tree/main/docs/cli`.

In addition to this CLI, the Tanzu portfolio includes a **graphical user interface** (**GUI**) tool named **Tanzu Mission Control** for all TKG cluster operations. We will learn about this tool in detail in *Chapter 9, Managing and Controlling Kubernetes Clusters with Tanzu Mission Control*.

Tanzu Kubernetes Grid installer

The TKG installer is a GUI component that provides a wizard for installing TKG on a selected infrastructure, which could be either vSphere, AWS, or Azure at the time of writing. During the initial setup of TKG, a very small bootstrapped Kubernetes cluster is deployed on the operator's workstation (also called a **bootstrap machine**). The TKG installer pods get deployed on that local Kubernetes cluster to deploy and start the GUI in the bootstrap machine. The operator then uses the locally running TKG installer's GUI to deploy the actual TKG foundation on the targeted cloud infrastructure. It seems to be confusing because here, TKG uses a (small) Kubernetes platform to deploy a (large) Kubernetes platform. We will use this GUI during our TKG installation steps in the next section.

It is worth noting that TKG also provides a way to install the foundation using the `tanzu` CLI as well, along with the GUI experience. However, it is recommended to use the GUI for the first installation as the wizard that's used in the GUI generates an installation configuration file that can later be used with little modifications to quickly install other similar TKG foundations using the `tanzu` CLI.

kind

kind stands for **Kubernetes inside Docker**. kind is an open source project under the umbrella of Kubernetes **Special Interest Groups** (**SIGs**) (`https://kind.sigs.k8s.io/`) that allows you to deploy a very tiny Kubernetes cluster as a container running inside a Docker environment. kind is a CNCF-conformant Kubernetes installer and typically gets deployed in local desktop environments to deploy a small Kubernetes cluster. The TKG installer pods get deployed in a kind cluster for bootstrap purposes only. Once the actual TKG foundation has been installed, the kind cluster is destroyed, along with the running installer components, since their purpose has come to an end.

Cluster API

Kubernetes includes a tool named **Kubeadm** (`https://kubernetes.io/docs/reference/setup-tools/kubeadm/`) that helps configure a server to make it a Kubernetes cluster node. It does this by installing the required Kubernetes-specific components. Kubeadm helps the node join the cluster as best practice *fast paths* for creating Kubernetes clusters. However, Kubeadm does not know how to provision the required infrastructure for the given cloud provider. Kubeadm requires the servers to be created with the required compute, storage, and networking setup before it can convert those servers into Kubernetes nodes. This gap in infrastructure management in Kubeadm is filled by the **Cluster API**.

A TKG user would never directly use the Cluster API, but it is one of the most important building blocks of TKG that interfaces with and abstracts the underlying cloud infrastructure. The term **API** stands for **application program interface** – a well-known term in the field of computer programming.

The Cluster API is also a SIG project (`https://cluster-api.sigs.k8s.io/`). The purpose of this project is to provide an interface layer to platforms such as TKG to perform various Kubernetes life cycle operations, including cluster provisioning, upgrading, and operating. As we know, different cloud providers have different ways of operating their infrastructure. A virtual machine in the world of vSphere is called an EC2 machine in the world of AWS. A virtual network boundary in AWS is known as a **Virtual Private Cloud** (**VPC**) but it is called a **Virtual Network** (**VNet**) in Azure. The implementation of the Cluster API interfaces abstracts such cloud-specific terminologies and operational differences.

To provide fully automated Kubernetes cluster life cycle management, TKG uses the Cluster API; the Cluster API uses Kubeadm internally. This way, TKG fully leverages different open source projects to provide a uniform multi-cloud Kubernetes cluster management experience.

Carvel

Carvel (`https://carvel.dev/`) is one additional open source package management toolkit that TKG uses for installing itself and other optional packages. Carvel is a very powerful and flexible packaging tool for Kubernetes deployments. Carvel contains multiple tools for different package management and deployment tasks, as listed in the following points:

- *kapp-controller*: To provide continuous delivery for apps and packages deployed on Kubernetes clusters in a GitOps way

- *ytt*: To create and use templatized YAML configurations for package deployment, allowing package configuration customization during their installations on Kubernetes clusters

- *kapp*: To bundle multiple Kubernetes resources as one application package that can be installed, upgraded, or deleted as one unit

- *kbld*: To build or reference container images used in the Kubernetes resource configuration in an immutable way

- *imgpkg*: To package, distribute, and relocate Kubernetes configuration and the associated container images as one bundle

- *vendir*: To declaratively state a directory's contents

- *secretgen-controller*: To manage various types of secrets used by the packages

After learning about the interface related components of TKG, let's review the core components of TKG now.

Core components of Tanzu Kubernetes Grid

In addition to the interfacing components that we saw previously, TKG contains a set of core components that are the most basic building blocks for a Kubernetes platform. Let's learn about them here.

Operating systems

One of the many advantages of using TKG is that you get a supported and hardened operating system from VMware for Kubernetes nodes. TKG supports Photon and Ubuntu Linux-based operating systems. Additionally, TKG allows you to build among the supported and a few other flavors of Linux and Windows operating systems with a certain level of customization.

Kubernetes

Kubernetes is the main ingredient of TKG. Each version of TKG includes one or more versions of upstream open source Kubernetes distributions without any modifications. This component is the main Kubernetes platform, including the main set of tools that make the platform. It includes the following tools:

- *kube-apiserver*: An API interface for the Kubernetes control plane
- *etcd*: A key-value store to persist the state of the Kubernetes cluster and its workload configuration
- *kube-scheduler*: To host newly created pods in a node based on various selection criteria
- *kube-controller-manager*: To run different control processes to manage nodes, jobs, service endpoints, and service accounts
- *cloud-controller-manager*: To run different cloud/infrastructure-specific processes to manage node health checks, routes, and cloud-specific service load balancers
- *kubelet*: A node-residing agent to ensure pods' running status and report back to the control plane
- *kube-proxy*: A node-residing network proxy that ensures service routes to their endpoint pods running on the node
- *Container runtime*: A node-residing software (**containerd** in the case of TKG) that runs and manages containers

In addition to these core components of Kubernetes, there are a few more core components that TKG includes that are required for the platform to operate. Let's take a look.

Metrics Server

Metrics Server aggregates resource usage data, such as container, node CPU, and memory usage, in a Kubernetes cluster and makes it available via the Metrics API defined in Kubernetes. This component is required to pull details after using the `kubectl top node` or `kubectl top pod` command.

Container Storage Interface (CSI)

Container Storage Interface (CSI) is a Kubernetes specification that requires implementation from the storage infrastructure provider. This will be used in the Kubernetes platform to provide persistent volumes for the workloads that need them. It provides Kubernetes users with more options for different storage solutions. One Kubernetes cluster may have different types of storage, including **solid-state drives (SSDs)**, **magnetic drives (HDDs)**, and other variants that provide different rates of data input and output. TKG uses the infrastructure-specific storage driver for vSphere, AWS, or Azure.

CoreDNS

CoreDNS (`https://coredns.io/`) is an open source DNS server that is used to provide Kubernetes service name resolution. The open source Kubernetes installs kube-dns for this purpose but allows you to replace kube-dns with CoreDNS, which is a more enhanced DNS server. TKG clusters get deployed with CoreDNS.

Container Network Interface (CNI)

As per the Kubernetes networking specification, every pod in a cluster should have a unique IP address and should be able to communicate with other pods on any other node of the cluster without any **network address translation** (**NAT**). Additionally, all the agents running in nodes, such as kubelet, should be able to communicate with each pod running on their respective node. These requirements ensure smooth communication between apps deployed on the same cluster. However, we need a specific networking arrangement to implement this specification. This is the CNI implementation, which is also known as the overlay network for Kubernetes clusters. There are many CNI implementations that we can choose from to be used in Kubernetes clusters. Out of them, TKG supports **Antrea** (`https://antrea.io/`) and **Calico** (`https://www.tigera.io/project-calico/`), which we can choose from during platform setup.

Control plane load balancers

TKG can create multi-master Kubernetes clusters for high availability of the control plane objects of a Kubernetes cluster. Such clusters typically have three Kubernetes control plane (master) nodes serving Kubernetes API traffic and performing crucial workload management activities. For a TKG deployment on AWS and Azure, TKG creates their respective virtual load balancer objects to front these control plane nodes for the API server traffic distribution. For vSphere, TKG includes **NSX Advanced Load Balancer** (`https://www.vmware.com/products/nsx-advanced-load-balancer.html`) to create a virtual load balancer for the same purpose. However, if TKG is not configured with NSX Advanced Load Balancer on vSphere, it uses an open source and lightweight virtual load balancer named **kube-vip** (`https://kube-vip.io/`). kube-vip is also a CNCF-governed project.

Extension packages of Tanzu Kubernetes Grid

The open source Kubernetes distribution comes with the minimal components required to deploy a Kubernetes platform. However, to deploy a production-grade Kubernetes environment, we need several other capabilities, such as logging, monitoring, access control, backup and restore, and more. Since TKG is an enterprise-grade Kubernetes distribution, it also comes with many such open source extension packages with VMware-signed binaries. Let's check these components out.

Logging

Fluent Bit (`https://fluentbit.io/`) is a high-performance-focused open source log forwarding tool for different flavors of Linux and Windows operating systems. Fluent Bit is also a CNCF sub-project. The purpose of this tool is to process logs emitted from Kubernetes nodes and the workload pods and can be configured to plumb them to a long list of possible log aggregation destinations, including Splunk, Elasticsearch, Amazon CloudWatch, Kafka, and many more.

Ingress controller

TKG supplies **Contour** (`https://projectcontour.io/`) binaries as an extended package to provide an ingress type of routing for the external-facing services deployed on Kubernetes clusters. Contour is an open source project under the CNCF umbrella that internally uses the **Envoy proxy** (`https://www.envoyproxy.io/`), another CNCF open source project. Together with Envoy (as the data plane), Contour (as the control plane) provides the required implementation of the ingress controller to provide the HTTP-level (network layer 7) service routing defined using the ingress resources of Kubernetes.

Identity and authentication

TKG includes **Pinniped** (`https://pinniped.dev/`), another CNCF open source project backed by VMware to provide an easy button for the Kubernetes cluster's user identity and authentication management. The upstream Kubernetes distribution does not include any authentication mechanism and only provides the configuration for authorization. Hence, to allow cluster users to get authenticated using existing identity providers based on a **Lightweight Directory Access Protocol (LDAP)** server or **Open ID Connect (OIDC)**, TKG supplies Pinniped as an extension package.

Observability

For cluster observability, TKG also supplies the signed binaries for **Prometheus** (`https://prometheus.io/`) and **Grafana** (`https://github.com/grafana/grafana`), two very popular open source monitoring tools for the Kubernetes ecosystem. Here, Prometheus is a metrics aggregator engine and Grafana is a metrics rendering tool for visualization. In addition to these *batteries included* monitoring tools, TKG also supports first-class integration with VMware Aria operations for Applications, a **Software-as-a-Service (SaaS)** platform in the VMware portfolio, for more capabilities around scaling, functionality, and power for full stack observability. We will cover this product in detail in *Chapter 10, Realizing Full-Stack Visibility with VMware Aria Operations for Applications*.

Container registry

TKG also comes with **Harbor**, a purpose-built container registry, as an extended package that can be installed in the cluster if required. We covered Harbor extensively in *Chapter 6, Managing Container Images with Harbor*.

Backup and restore

Disaster recovery is a very important aspect of platform-supporting production applications. A Kubernetes cluster is not any different running critical production workloads. However, Kubernetes does not include anything to back up and restore the state of the workloads running on its clusters. To fill this gap, TKG includes **Velero** (`https://velero.io/`) as an extension package in the bundle. Velero is also an open source project under the CNCF umbrella. Velero provides ways to take backups and restore them later at the cluster level, Kubernetes namespace level, or for specific workloads based on their attached metadata. Velero can also take backups of the persistent volumes containing stateful application data and restore them when required. This is the tool that is used under the hood of Tanzu Mission Control for backup and recovery features. We will cover this in detail in *Chapter 9, Managing and Controlling Kubernetes Clusters with Tanzu Mission Control*.

ExternalDNS

TKG also supplies **ExternalDNS** (`https://github.com/kubernetes-sigs/external-dns`), an open source Kubernetes SIG project, as an extension package. ExternalDNS allows you to control the DNS records dynamically for the external-facing services deployed on the cluster using a Kubernetes resource definition file in a way that is agnostic to a DNS provider. The external services running on a Kubernetes cluster can get the required DNS record binding handled by ExternalDNS on a linked DNS server such as AWS Route53 or Google Cloud DNS. In a way, it provides a way to control external DNS configurations using Kubernetes resource definitions.

cert-manager

cert-manager (`https://cert-manager.io/`) is another CNCF open source project that TKG includes as an extension to manage X.509 (identity) certificates used in a Kubernetes cluster. cert-manager obtains certificates from a variety of configured certificate issuers, ensures certificate validity, and tries to renew them before expiry.

Now that we have seen what components TKG contains, let's learn about some of the important concepts of this platform.

Important concepts of Tanzu Kubernetes Grid

TKG is a distributed system with several moving parts. To understand how TKG works, we need to learn a few concepts of this system. Let's take a look.

Bootstrap cluster

As discussed earlier in this chapter, TKG uses a kind cluster to deploy a TKG foundation on the selected infrastructure. This kind cluster is very small and runs in a Docker container in the operator's workstation, which is typically a personal computer. This kind cluster contains the required machinery, including the Tanzu installation portal and other components that help bootstrap a TKG foundation. Because of this, this kind cluster is also known as a bootstrap cluster.

Tanzu Kubernetes releases (TKrs)

A TKG deployment may support multiple different versions of Kubernetes. A TKr is a custom resource definition under TKG that contains a reference to one such Kubernetes version that TKG can deploy and manage. TKrs include components such as Antrea with its linked version definition for the given Kubernetes version. The management cluster of TKG runs a TKr controller that keeps checking the public registry for a new Kubernetes version availability. Once a new version is available, the TKr controller downloads the required artifacts on the TKG management cluster to make it available for use. This way, one TKG management cluster may deploy and manage multiple versions of Kubernetes clusters supported under that TKG version. This arrangement provides flexibility to different teams wanting to run their applications on different Kubernetes versions that are still managed by the same TKG control plane.

Management cluster

Every TKG foundation has one management cluster. A management cluster is nothing but a Kubernetes cluster running specific workloads used to life cycle other Kubernetes clusters. A TKG foundation and thus its management cluster is infrastructure specific. Because of that, we need one management cluster for vSphere, one for AWS, and one for Azure if we want to deploy TKG clusters on all the platforms. This is because a management cluster contains underlying cloud-specific Cluster API configuration. This way, a management cluster can create multiple Kubernetes workload clusters on the same cloud platform where it is deployed. In addition to creating, scaling, upgrading, and deleting a Kubernetes cluster, the management cluster also keeps track of the actual versus the desired state of the Kubernetes cluster nodes if it is configured to do so. The management cluster restarts or recreates an unhealthy or missing node from a cluster it manages. A management cluster is just a normal Kubernetes cluster and can also run any custom app, but it should only be used for its main purpose, which is to operate a large number of Kubernetes clusters. Its access and permission should be very much restrictive to the TKG platform operations team considering the level of access it has over other Kubernetes clusters managed by it. The TKG platform operators may give limited access to a management cluster using Kubernetes namespaces under the management cluster. This way, different teams can use the same management cluster to life cycle their Kubernetes clusters linked with a specific namespace of the management cluster. In addition to the upstream Kubernetes components, a TKG management cluster has the following TKG-specific components deployed into it:

- Cluster API components
- cert-manager

- secretgen-controller
- kapp-controller
- tkr-controller

Workload cluster

A workload cluster is a normal Kubernetes cluster created by a management cluster. A workload cluster is where we deploy our apps. Depending on the organization's practices and scale, different teams and their application environments may use separate workload clusters. The size in terms of the number of nodes in a workload cluster is only limited to the infrastructure availability. However, it is recommended to keep a cluster size as small as possible to reduce the blast radius if something goes wrong and for quicker maintenance time. TKG makes it very easy to manage many Kubernetes clusters. As we discussed previously, workload clusters under a management cluster may have different supported versions of Kubernetes based on the requirements of the teams using them.

Node pool

A TKG cluster may have different types of worker nodes that have different configurations and resources attached to a cluster. Such heterogeneous node types allow different kinds of workloads to leverage them for specific resource requirements or just for isolation purposes. For example, a workload cluster may have some number of nodes with a consumable **graphical processing unit (GPU)** that can be utilized by extremely compute-hungry machine learning workloads deployed on the cluster. We can add such different types of nodes to a workload cluster using **node pools** in TKG. Later, we can configure such nodes with taints and tolerations to only allow the workloads that need to use certain types of nodes to be scheduled.

Deployment topologies

TKG supports two different deployment topologies for creating the management and workload clusters:

1. **Dev Plan**: In this topology, TKG creates a single control plane node and the required number of worker nodes. This plan is used for non-critical deployments that can tolerate the unavailability of the Kubernetes API server, etcd, and other control plane functions. This topology requires fewer resources and is typically used for lab environments.

2. **Prod Plan**: In this topology, TKG creates three control plane nodes and fronts them with a load balancer to provide a highly available Kubernetes control plane. As the name suggests, it is used for clusters that will host critical workloads and may not tolerate any control plane downtime for cluster operations.

Figure 7.3 shows how TKG works at a high level. As shown in this figure, an operator uses either the `tanzu` CLI or the TKG bootstrap UI to supply the required configuration for the foundation. Once the management cluster has been created, the operator can directly use it to create the required number of workload clusters:

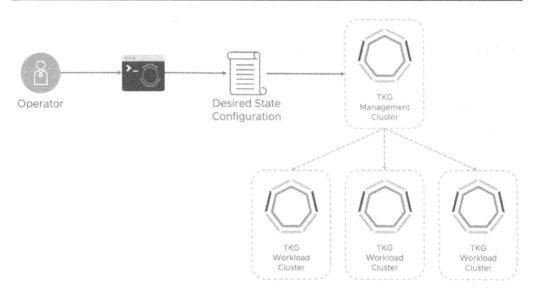

Figure 7.3 – TKG layout

In the next section, we will learn how to create a TKG management cluster on AWS and understand the operation flow in detail.

Getting started with Tanzu Kubernetes Grid

TKG, being a multi-cloud solution, can be installed on a vSphere-based on-premises environment, or Microsoft Azure and **Amazon Web Services** (**AWS**)-based public cloud platforms. To keep this chapter to an acceptable length, we will only cover how to install and configure a TKG foundation on AWS. You may find additional TKG installation and configuration details here: `https://docs.vmware.com/en/VMware-Tanzu-Kubernetes-Grid/1.5/vmware-tanzu-kubernetes-grid-15/GUID-mgmt-clusters-prepare-deployment.html`.

We will perform the following tasks in this section:

- Configure the bootstrap machine – the operator workstation from where the installation will be triggered
- Deploy the TKG management cluster
- Create a TKG workload cluster using the management cluster
- Obtain access to the workload cluster

But before we do that, we need to ensure that the following prerequisites are met to complete these tasks.

Prerequisites

The following are the prerequisites to follow the TKG installation instructions given in this section:

- An AWS account with the following:

  - An access key and an access key secret

  - An SSH key pair registered with the account for the region where TKG is being installed

  - Permission to create a CloudFormation stack that defines **Identity and Access Management (IAM)** resources and their permissions

  - A sufficient resource quota that's allowed to create two **Virtual Private Clouds (VPCs)**, nine subnets (two internet-facing and one internal per availability zone) in the VPC, four EC2 security groups, two internet gateways, three NAT gateways, and three Elastic IP addresses in the selected region for TKG deployment

- A Linux or Mac workstation with the following:

  - Internet access

  - Command-line access

  - Web browser access

  - Port 6443 access to all the EC2 instances to access Kubernetes APIs

  - Docker Desktop installed and running with 6 GB allocated

  - 2-core CPU

  - The kubectl CLI v1.22 or higher

- Access to a **Network Time Protocol (NTP)** server

- Access to https://my.vmware.com/ with an account set up to download the required binaries

Let's start with the first task, which is to prepare the bootstrap machine that will be used for this installation. We will need a few tools and environmental configurations before we begin the installation.

Configuring the bootstrap machine

The following sub-tasks prepare a bootstrap machine with the required tools and configuration for TKG setup on AWS.

Installing Tanzu and the Kubectl CLI

Follow these steps:

1. Create a directory on your local machine where you will store the required artifacts:

    ```
    $ mkdir $HOME/tkg-154
    ```

2. Go to https://my.vmware.com/ and log in using your My VMware credentials.

3. Go to https://customerconnect.vmware.com/downloads/
    details?downloadGroup=TKG-154&productId=1162 to download the
    required artifacts.

4. Make sure that the selected version in the dropdown is **1.5.4**:

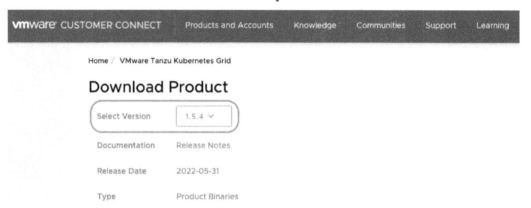

Figure 7.4 – Selecting a download version

5. Under the **Product Downloads** tab, scroll to the **VMware Tanzu CLI 1.5.4** section and download
    the binary for your operating system. Note that the procedure followed in this chapter is being
    done on a macOS machine. While most of the commands listed in this chapter should work
    on the other platforms, there could be some minor differences:

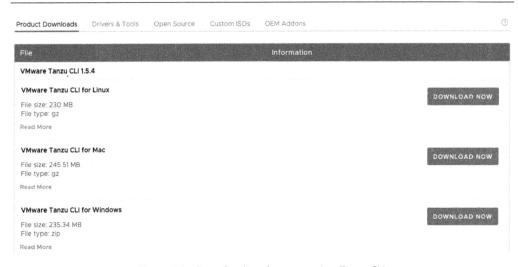

Figure 7.5 – Downloading the appropriate Tanzu CLI

6. Accept the End User License Agreement:

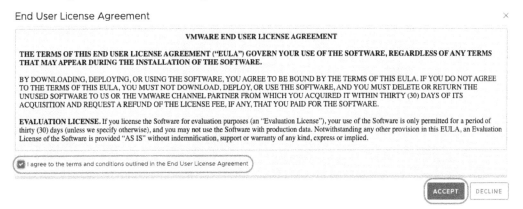

Figure 7.6 – Accepting the End User License Agreement

7. Download the binary into the $HOME/tkg-154 directory that we created in *step 1*.

8. Now, on the same page when you downloaded the binaries, go to the **Kubectl 1.22.9 for VMware Tanzu Kubernetes Grid 1.5.4** section:

Kubectl 1.22.9 for VMware Tanzu Kubernetes Grid 1.5.4

Kubectl cluster cli v1.22.9 for Linux

File size: 12.9 MB
File type: gz

Read More

DOWNLOAD NOW

Kubectl cluster cli v1.22.9 for Mac

File size: 14.13 MB
File type: gz

Read More

DOWNLOAD NOW

Kubectl cluster cli v1.22.9 for Windows

File size: 13.13 MB
File type: gz

Read More

DOWNLOAD NOW

Figure 7.7 – Downloading the appropriate Kubectl CLI

9. Download the binary into the same $HOME/tkg-154 directory.

10. Go into the $HOME/tkg-154 directory and extract the CLI binaries you downloaded previously. Run the following commands to do this:

```
$ cd $HOME/tkg-154
$ tar -xvf tanzu-cli-bundle-darwin-amd64.tar.gz
$ gunzip -dvf kubectl-mac-v1.22.9+vmware.1.gz
```

11. Install the Tanzu CLI on you local system:

```
$ sudo install cli/core/v0.11.6/tanzu-core-darwin_amd64 /
usr/local/bin/tanzu
```

12. Verify the installation by running the following command:

```
$ tanzu version
```

You should see version 0.11.6 in the output.

13. Initialize the Tanzu CLI:

```
$ tanzu init
```

14. Clean up any pre-existing Tanzu plugins for a clean start:

```
$ tanzu plugin clean
```

15. Make sure you are under the $HOME/tkg-154 directory, which contains the extracted cli directory:

```
$ cd $HOME/tkg-154/
```

16. Run the following command to install all the required plugins for this TKG release:

```
$ tanzu plugin sync
```

You should be able to see the following output for the command's execution:

```
Checking for required plugins...
Installing plugin 'cluster:v0.11.6'
Installing plugin 'kubernetes-release:v0.11.6'
Installing plugin 'login:v0.11.6'
Installing plugin 'management-cluster:v0.11.6'
Installing plugin 'package:v0.11.6'
Installing plugin 'pinniped-auth:v0.11.6'
Installing plugin 'secret:v0.11.6'
Successfully installed all required plugins
⮡ Done
```

17. Verify the plugin's installation status by running the following command:

```
$ tanzu plugin list
```

You should be able to see all the plugins listed, along with their versions and statuses, as shown in the following screenshot:

```
~ ⬥ tanzu plugin list
NAME                 DESCRIPTION                                                   SCOPE       DISCOVERY                       VERSION  STATUS
cluster              Kubernetes cluster operations                                 Context     default-tkg-aws-mgmt-cluster    v0.11.6  installed
kubernetes-release   Kubernetes release operations                                 Context     default-tkg-aws-mgmt-cluster    v0.11.6  installed
login                Login to the platform                                         Standalone  default                         v0.11.6  installed
management-cluster   Kubernetes management-cluster operations                      Standalone  default                         v0.11.6  installed
package              Tanzu package management                                      Standalone  default                         v0.11.6  installed
pinniped-auth        Pinniped authentication operations (usually not directly invoked)  Standalone  default                    v0.11.6  installed
secret               Tanzu secret management                                       Standalone  default                         v0.11.6  installed
```

Figure 7.8 – Installed TKG plugin list

Now, let's install the Kubectl CLI.

18. Run the following commands from the $HOME/tkg-154 directory, which is where we downloaded and extracted the Kubectl CLI:

```
$ chmod ugo+x kubectl-mac-v1.22.9+vmware.1
$ sudo install kubectl-mac-v1.22.9+vmware.1 /usr/local/
bin/kubectl
```

19. Verify the installation by running the kubectl version command. You should see the client version as v1.22.9+vmware.1.

Installing Carvel tools

As we discussed in the previous section, TKG uses the Carvel toolkit for its packaging and installation. For that reason, we will need some of the Carvel tool's CLI binaries installed in the bootstrap machine. The Tanzu CLI bundle that we previously downloaded and extracted contains all these tools. The following steps describe the procedure to install them:

1. Go to the cli directory under $HOME/tkg-154:

```
$ cd $HOME/tkg-154/cli
```

2. Install **ytt** with the following commands:

```
$ gunzip ytt-darwin-amd64-v0.37.0+vmware.1.gz
$ chmod ugo+x ytt-darwin-amd64-v0.37.0+vmware.1
$ mv ./ytt-darwin-amd64-v0.37.0+vmware.1 /usr/local/bin/
ytt
```

3. Verify the **ytt** installation by running the ytt --version command.

4. Install **kapp** with the following commands:

```
$ gunzip kapp-darwin-amd64-v0.42.0+vmware.2.gz
$ chmod ugo+x kapp-darwin-amd64-v0.42.0+vmware.2
$ mv ./kapp-darwin-amd64-v0.42.0+vmware.2 /usr/local/bin/
kapp
```

5. Verify the **kapp** installation by running the kapp --version command.

6. Install **kbld** with the following commands:

```
$ gunzip kbld-darwin-amd64-v0.31.0+vmware.1.gz
$ chmod ugo+x kbld-darwin-amd64-v0.31.0+vmware.1
$ mv ./kbld-darwin-amd64-v0.31.0+vmware.1 /usr/local/bin/
kbld
```

7. Verify the **kbld** installation by running the `kbld --version` command.

8. Install **imgpkg** with the following commands:

    ```
    $ gunzip imgpkg-darwin-amd64-v0.22.0+vmware.1.gz
    $ chmod ugo+x imgpkg-darwin-amd64-v0.22.0+vmware.1
    $ mv ./imgpkg-darwin-amd64-v0.22.0+vmware.1 /usr/local/
    bin/imgpkg
    ```

9. Verify the **imgpkg** installation by running the `imgpkg --version` command.

Installing AWS-specific tools

Deploying a TKG foundation on the AWS platform requires the **aws** CLI to be installed on the bootstrap machine. Let's configure this:

1. Install the **aws** CLI using the instructions provided at `https://docs.aws.amazon.com/cli/latest/userguide/getting-started-install.html`.

2. Verify the installation of the **aws** CLI by running the following command:

    ```
    $ aws -version
    ```

    You should be able to see the CLI version listed, as shown in the following output. The version could be different in your case, depending on when it is installed, but it should be v2.x:

    ```
    aws-cli/2.5.3 Python/3.9.11 Darwin/21.4.0 exe/x86_64
    prompt/off
    ```

3. Run the following command to configure the access profile for your AWS account to be used for this installation with the previously defined permissions and quotas:

    ```
    $ aws configure -profile tkg
    ```

4. Supply the values of the access key, secret access key, region, and command output format, as shown in the following code:

    ```
    AWS Access Key ID [None]: ********************
    AWS Secret Access Key [None]: **************************
    *************
    Default region name [None]: us-east-1
    Default output format [None]: text
    ```

    Here, you may replace `us-east-1` with any other AWS region of your choice with the previously listed prerequisites fulfilled.

5. Run the following command to ensure you can see the existing SSH key pair in the region as it was listed in the prerequisites:

```
$ aws ec2 describe-key-pairs --profile tkg
```

This command's output should show at least one key pair listed.

We now have all the required tools configured in the bootstrap machine to begin installing the TKG management cluster. Let's begin.

Installing the management cluster

In this section, we will use the Tanzu installer UI to deploy the management cluster on the configured AWS account.

> **Important note**
>
> By default, the upstream Kubernetes distribution does not come with user authentication capabilities and allows the user to have admin-level access. To fill this gap, TKG comes with Pinniped to integrate an external LDAP or OIDC identity provider. To minimize installation prerequisites and complexity, we will not use such an integration, which requires a pre-existing LDAP or OIDC setup access. A real-life TKG cluster should never be configured without such integration with an external identity provider. You can learn more about this topic here: https://docs.vmware.com/en/VMware-Tanzu-Kubernetes-Grid/1.5/vmware-tanzu-kubernetes-grid-15/GUID-iam-configure-id-mgmt.html.
>
> Additionally, despite aiming for minimal complexity and an infrastructure footprint, following this **TKG configuration on AWS will incur cloud service usage charges in your AWS account** as it will use the EC2 instance types that are not qualified for the free-tier credits, along with some other chargeable services such as Elastic IP, NAT gateways, EBS volumes, Elastic Load Balancers, and a few others. If you plan to follow this guide to install TKG on AWS, it is recommended that you also clean up the resources using the procedure described later in this chapter, followed by a manual inspection to verify the removal of all provisioned resources.

The following steps outline the installation procedure:

1. Ensure the local Docker daemon is running to deploy and run containers on the bootstrap machine.
2. Run the following command to start the installer UI in the default browser window of the bootstrap workstation. This UI supports Chrome, Safari, Firefox, Internet Explorer, and Edge, along with their considerably newer versions:

```
$ tanzu mc create --ui
```

This command should automatically open the browser window with the UI running. Otherwise, it can be accessed using http://127.0.0.1:8080/.

3. Click on the **DEPLOY** button under **Amazon Web Services**, as highlighted in the following screenshot:

Figure 7.9 – Selecting Amazon Web Services for deployment

4. Select the necessary AWS account details by following these sub-steps; these are highlighted in the following screenshot:

I.   Set **AWS CREDENTIAL TYPE** to **Credential Profile**.

II.   Select **tkg** from the **AWS CREDENTIAL PROFILE** dropdown and its associated **REGION** that we configured during the **aws** CLI setup earlier in this chapter.

III. Click the **CONNECT** button to ensure the UI can connect to the select AWS account using the previously created configuration.

IV.   Click the **NEXT** button to move to the next configuration section:

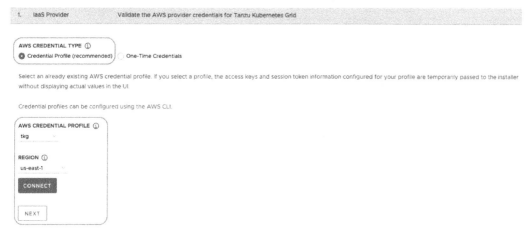

Figure 7.10 – Selecting an AWS account for the installation

5. Click the **NEXT** button to choose the default VPC configuration, as shown in the following screenshot:

Figure 7.11 – Choosing the default VPC configuration

6. Set the management cluster deployment plan to **Development** and **INSTANCE TYPE** to **t3.large**. This instance type has 2 vCPUs and 8 GiB memory, which is just good enough for a lab-like setup:

Figure 7.12 – Selecting the management cluster's deployment plan and instance type

7. Enter other details of the management cluster, as listed in the following sub-steps and shown in the following screenshot:

    I.   Enter a short name under **MANAGEMENT CLUSTER NAME** – that is, `tkg-aws-mgmt-cluster`.

    II.   Enter the name of the SSH key-pair that's available to the account you obtained in *step 5* of the **aws** CLI configuration under **EC2 KEY PAIR**.

    III.   Leave the default selection as-is for the checkbox options.

    IV.   Select one of the availability zones from the **AVAILABILITY ZONE 1** dropdown for the selected region to deploy the management cluster into. For the **Production** deployment plan, we need to select three availability zones for the three control plane nodes of the management cluster.

V.   Select **t3.large** as the worker node instance type from the **AZ1 WORKER NODE INSTANCE TYPE** dropdown.

VI.   Click the **NEXT** button to move on:

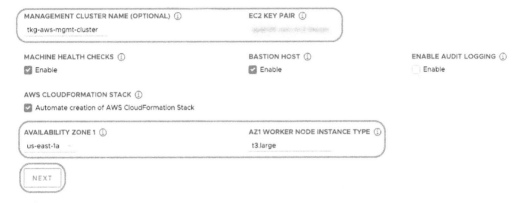

Figure 7.13 – Entering the management cluster's details

8.   Optionally, enter management cluster metadata and click the **NEXT** button:

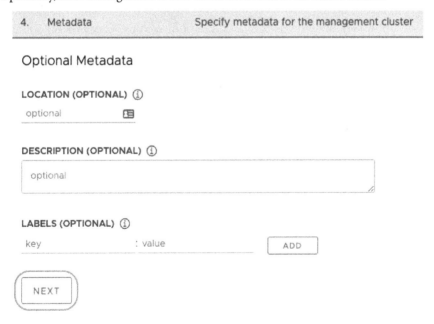

Figure 7.14 – Entering the management cluster's metadata

9.   Leave the default Kubernetes container network configuration as-is and click the **NEXT** button:

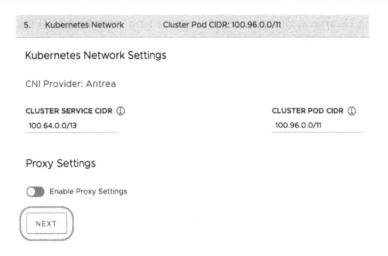

Figure 7.15 – Leaving the default Kubernetes network configuration as-is

10. Disable the identity management settings and click the **NEXT** button:

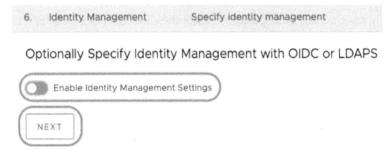

Figure 7.16 – Disabling identity management settings

11. Select the management cluster's operating system image, as highlighted in the screenshot, and click the **NEXT** button:

Figure 7.17 – Selecting the management cluster's operating system

12. Optionally, choose to participate in VMware's **Customer Experience Improvement Program (CEIP)** and click the **NEXT** button:

Figure 7.18 – Choosing to participate in the Customer Experience Improvement Program

13. Click the **REVIEW CONFIGURATION** button to verify the inputs before triggering the management cluster creation on AWS:

← 🔘 Deploy Management Cluster on Amazon Web Services

>	⊘	IaaS Provider	Validate the AWS provider credentials for Tanzu Kubernetes Grid
>	⊘	VPC for AWS	Specify VPC settings for AWS
>	⊘	Management Cluster Settings	Development cluster selected: 1 node control plane
>	⊘	Metadata	Specify metadata for the management cluster
>	⊘	Kubernetes Network	Cluster Pod CIDR: 100.96.0.0/11
>	⊘	Identity Management	Specify identity management
>	⊘	OS Image	OS Image: amazon-2-amd64 (ami-0092b98bc500b4c18)
>	⊘	CEIP Agreement	Join the CEIP Program for TKG

REVIEW CONFIGURATION

Figure 7.19 – Opening the configuration summary for review

14. The following screenshot shows the bottom of the verification summary page. The command that's displayed can be used in the future to trigger the same deployment again using the Tanzu CLI. The **EXPORT CONFIGURATION** link at the bottom-right corner allows us to export this configuration in a file that can be used as a reference to deploy other management clusters on AWS with the required modifications. Finally, click the **DEPLOY MANAGEMENT CLUSTER** button to trigger the deployment:

Figure 7.20 – Triggering the management cluster's deployment

15. You will see various deployment logs and their installation statuses, as shown in the following screenshot. As you can see in the logs, the Cluster API is creating the required infrastructure component to deploy the management cluster on AWS:

Figure 7.21 – Deployment status with logs

16. The installation should ideally be completed in about 10 to 15 minutes and you should see a success message, as shown in the following screenshot. As highlighted in this screenshot, the logs also highlight how to access the management cluster from the bootstrap machine to perform different TKG operations. You should also be able to see these logs on the command line from where you fired the `tanzu mc create --ui` command, which brought up the installation browser window:

Figure 7.22 – Successful management cluster installation

Now that we have finished installing the management cluster, let's learn how to use it to create our first TKG workload cluster.

Creating a workload cluster

The following steps outline the procedure to access the newly created TKG management cluster using the bootstrap machine from where we triggered the management cluster installation. As a part of the installation steps, TKG adds the kubeconfig details to the bootstrap machine to allow administrator-level access to the management cluster and hence the Tanzu CLI pointing to the management cluster. Let's use the management cluster and the Tanzu CLI to create our first workload cluster. The following steps will be performed on the bootstrap machine:

1. Run the following command to set the kubectl context pointing to the newly created TKG management cluster:

    ```
    $ kubectl config use-context tkg-aws-mgmt-cluster-admin@
    tkg-aws-mgmt-cluster
    ```

2. Run the tanzu mc get command to view the details of the management cluster. The command will show that the management cluster has one control plane and one worker node created:

```
~ » tanzu mc get

NAME                      NAMESPACE      STATUS    CONTROLPLANE    WORKERS   KUBERNETES          ROLES        PLAN
tkg-aws-mgmt-cluster      tkg-system     running   1/1             1/1       v1.22.9+vmware.1    management   dev

Details:

NAME                                                                            READY  SEVERITY  REASON  SINCE  MESSAGE
/tkg-aws-mgmt-cluster                                                           True                      45m
 ├─ClusterInfrastructure - AWSCluster/tkg-aws-mgmt-cluster                      True                      46m
 ├─ControlPlane - KubeadmControlPlane/tkg-aws-mgmt-cluster-control-plane        True                      45m
 │ └─Machine/tkg-aws-mgmt-cluster-control-plane-dl44m                           True                      46m
 └─Workers
    └─MachineDeployment/tkg-aws-mgmt-cluster-md-0                               True                      46m
       └─Machine/tkg-aws-mgmt-cluster-md-0-7c8864bbcf-79d2v                     True                      46m

Providers:

NAMESPACE                            NAME                     TYPE                     PROVIDERNAME   VERSION   WATCHNAMESPACE
capa-system                          infrastructure-aws       InfrastructureProvider   aws            v1.2.0
capi-kubeadm-bootstrap-system        bootstrap-kubeadm        BootstrapProvider        kubeadm        v1.0.1
capi-kubeadm-control-plane-system    control-plane-kubeadm    ControlPlaneProvider     kubeadm        v1.0.1
capi-system                          cluster-api              CoreProvider             cluster-api    v1.0.1
```

Figure 7.23 – Management cluster details

To create a workload cluster, we need to supply the workload cluster configuration file to the management cluster. It contains the following details:

- Cluster name

- Cluster deployment plan (Development or Production)

- Worker node count

- Worker node EC2 type

- AWS-specific configurations such as region, AZ, network, and SSH key

- Node operating system

- Node health check configuration

- Node-level auto-scaling configuration

We do not need to include all the attributes in the configuration file and may specify only the required attributes. For a broader list of attributes for the configuration file, visit `https://docs.vmware.com/en/VMware-Tanzu-Kubernetes-Grid/1.5/vmware-tanzu-kubernetes-grid-15/GUID-tanzu-k8s-clusters-aws.html#tanzu-kubernetes-cluster-template-0`.

We will download a preconfigured file from this book's GitHub repository and use it to create a workload cluster.

3. Copy the workload cluster configuration file into the bootstrap machine using the following command:

```
$ curl https://raw.githubusercontent.com/PacktPublishing/
DevSecOps-in-Practice-with-VMware-Tanzu/main/chapter-
07/tkg-workload-cluster-config.yaml -o $HOME/tkg-154/
tkg-workload-cluster-config.yaml
```

4. The config file should be available in the `$HOME/tkg-154/` directory and be called `tkg-workload-cluster-config.yaml`. Open the file in your choice of editor and update the following parameter values:

    I.   Update **AWS_NODE_AZ, AWS_NODE_AZ_1** and **AWS_NODE_AZ_2** based on the selected region if you have used any AWS region other than **us-east-1**.

    II.   Update **AWS_REGION** as per your **AWS_PROFILE** configuration if applicable.

    III. Update **AWS_SSH_KEY_NAME** to use the SSH key in your AWS account in the selected region. This is a must-change.

5. Run the following command to create the workload cluster using the configuration file we prepared in the previous step. Here, we are creating the cluster with Kubernetes v1.21.11 using the `--tkr` option so that we can learn how to upgrade the cluster later to v.1.22.9:

```
$ tanzu cluster create --file $HOME/tkg-154/tkg-workload-
cluster-config.yaml --tkr v1.21.11---vmware.1-tkg.3
```

The workload cluster should be created in about 10 to 15 minutes if all the configuration changes are done correctly.

6. Verify the cluster's creation status using the following command:

```
$ tanzu cluster list
```

You should see that the cluster is running, as shown in the following screenshot:

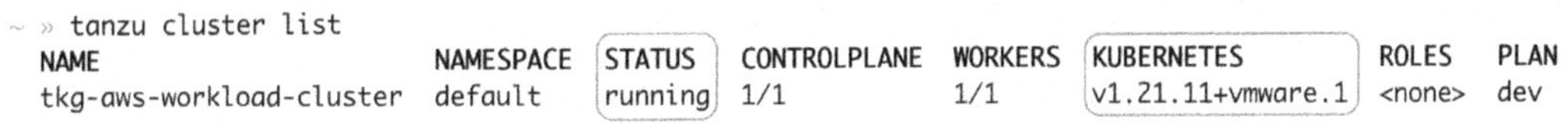

```
~ » tanzu cluster list
NAME                      NAMESPACE   STATUS    CONTROLPLANE   WORKERS   KUBERNETES          ROLES    PLAN
tkg-aws-workload-cluster  default     running   1/1            1/1       v1.21.11+vmware.1   <none>   dev
```

Figure 7.24 – Verifying the workload cluster's creation

With that, the TKG workload cluster has been created. Now, let's access the workload cluster.

7. Run the following command to obtain `kubeconfig` for the workload cluster:

```
$ tanzu cluster kubeconfig get tkg-aws-workload-
cluster  --admin
```

8. Run the following command to switch the kubectl context so that it points to the workload cluster:

```
$ kubectl config use-context tkg-aws-workload-cluster-admin@tkg-aws-workload-cluster
```

9. Run the following command to list the nodes of the workload cluster to ensure connectivity:

```
$ kubectl get nodes
```

You should be able to see the list of Kubernetes nodes, as shown in the following screenshot:

```
~ » kubectl get nodes
NAME                         STATUS   ROLES                  AGE   VERSION
ip-10-0-17-250.ec2.internal  Ready    <none>                 72m   v1.21.11+vmware.1
ip-10-0-20-55.ec2.internal   Ready    control-plane,master   73m   v1.21.11+vmware.1
```

Figure 7.25 – Verifying the workload cluster's connectivity

With this, we have completed all the tasks required to get started with TKG. We started with the bootstrap machine's configuration by installing all the required tools and CLIs. After that, we created a TKG management cluster on AWS using the Tanzu installer UI. And finally, we created a TKG workload cluster using the management cluster that we created. It is worth noting that one bootstrap machine may have a reference to more than one TKG management cluster. The operator may use different management clusters to manage the workload clusters under them by just switching the kubectl context to an appropriate management cluster config, followed by using the tanzu login command to get authenticated for the management cluster usage. Now, in the next and the last section of this chapter, we will learn about some of the most common day-2 activities around TKG.

Common day-2 operations with Tanzu Kubernetes Grid

Now that we have a fully configured and running TKG foundation on AWS, let's learn how to perform some of the day-2 operations on it. TKG makes these operations very trivial as they do all the heavy lifting behind the scenes:

- Scale a cluster to add or remove nodes
- Upgrade a cluster to bump up the Kubernetes version
- Delete a cluster
- Delete the entire TKG foundation

Let's start by scaling the workload cluster so that it has three worker nodes instead of just one.

Scaling a Tanzu Kubernetes Grid cluster

Run the following commands to scale the workload cluster we created:

1. Switch the kubectl context so that it's pointing to the management cluster that we previously created, which we used to create the workload cluster:

    ```
    $ kubectl config use-context tkg-aws-mgmt-cluster-admin@
    tkg-aws-mgmt-cluster
    ```

2. Ensure that the workload cluster has only one worker node by running the following command:

    ```
    $ tanzu cluster list
    ```

    You should see the following output, which shows 1/1 worker nodes:

    ```
    ~ » tanzu cluster list
    NAME                       NAMESPACE  STATUS   CONTROLPLANE  WORKERS  KUBERNETES        ROLES   PLAN
    tkg-aws-workload-cluster   default    running  1/1           1/1      v1.21.11+vmware.1  <none>  dev
    ```

    Figure 7.26 – Ensuring the worker node count

3. Run the following command to add two more worker nodes, creating the desired total count of three:

    ```
    $ tanzu cluster scale tkg-aws-workload-cluster -w 3
    ```

    You should see the following message, showing that the scaling is in progress:

    ```
    ~ » tanzu cluster scale tkg-aws-workload-cluster -w 3
    Successfully updated worker node machine deployment replica count for cluster tkg-aws-workload-cluster
    Workload cluster 'tkg-aws-workload-cluster' is being scaled
    ```

    Figure 7.27 – Worker node scaling in progress

4. Verify that the scaling has been done by running the following cluster listing command:

    ```
    $ tanzu cluster list
    ```

    You should now see 3/3 worker nodes in the output:

    ```
    ~ » tanzu cluster list
    NAME                       NAMESPACE  STATUS   CONTROLPLANE  WORKERS  KUBERNETES        ROLES   PLAN
    tkg-aws-workload-cluster   default    running  1/1           3/3      v1.21.11+vmware.1  <none>  dev
    ```

    Figure 7.28 – Confirming that the cluster scaling has been done

These steps showed you how to scale up a TKG cluster. The same procedure is also applicable to scale down a cluster. The -w option of the scale command declares the desired count of the worker nodes. And depending on the changes in the desired count, TKG adds or removes the worker nodes. The scaling command also has options to scale the control plane nodes or the nodes of a specific node pool. You can learn more about scaling by running the tanzu cluster scale --help command.

Upgrading a Tanzu Kubernetes Grid cluster

Now that we've scaled, let's learn how to upgrade the TKG workload cluster to deploy a newer version of Kubernetes. TKG allows such on-demand upgrades of selected clusters under a management cluster. Here, the owners of the cluster have the choice of which Kubernetes version they need so that they have enough time to prepare to upgrade their workload clusters. The following steps outline the upgrade procedure:

1. Switch the kubectl context so that it's pointing to the management cluster that we previously created, which we used to create the workload cluster:

   ```
   $ kubectl config use-context tkg-aws-mgmt-cluster-admin@
   tkg-aws-mgmt-cluster
   ```

2. Ensure that the workload cluster is deployed with Kubernetes v1.21.11 by using the following command:

   ```
   $ tanzu cluster list
   ```

   You should see the following output, showing that the cluster has been deployed with Kubernetes version 1.21.11:

```
~ » tanzu cluster list
NAME                        NAMESPACE  STATUS   CONTROLPLANE  WORKERS  KUBERNETES         ROLES    PLAN
tkg-aws-workload-cluster    default    running  1/1           3/3      v1.21.11+vmware.1  <none>   dev
```

Figure 7.29 – Checking the current Kubernetes version of the cluster

3. Run the following command to check the available Kubernetes version(s) for the upgrade:

   ```
   $ tanzu kubernetes-release get
   ```

   You should see the following output, which shows v1.22.9 as an option to upgrade v1.21.11, as highlighted in the following screenshot:

```
~ » tanzu kubernetes-release get
NAME                      VERSION                  COMPATIBLE  ACTIVE  UPDATES AVAILABLE
v1.20.15---vmware.1-tkg.2 v1.20.15+vmware.1-tkg.2  True        True    True
v1.21.11---vmware.1-tkg.3 v1.21.11+vmware.1-tkg.3  True        True    True
v1.22.9---vmware.1-tkg.1  v1.22.9+vmware.1-tkg.1   True        True    False
```

Figure 7.30 – Checking the available version upgrade options

4. Run the following command to upgrade the workload cluster to Kubernetes version 1.22.9:

   ```
   $ tanzu cluster upgrade tkg-aws-workload-cluster --tkr
   v1.22.9---vmware.1-tkg.1
   ```

The --tkr option mentions the target version for the upgrade that we picked from the available version list in the previous step. This command will upgrade the workload cluster in a rolling manner, one node at a time, to minimize workload downtime. The applications running with more than one pod would not face any downtime during such rolling upgrades. After firing the preceding command, you should see the following confirmation message in about 15 to 20 minutes:

```
~ » tanzu cluster upgrade tkg-aws-workload-cluster --tkr v1.22.9---vmware.1-tkg.1
Upgrading workload cluster 'tkg-aws-workload-cluster' to kubernetes version 'v1.22.9+vmware.1'. Are you sure? [y/N]: y
Validating configuration...
unable to create AWS client. Skipping validations that require an AWS client
updating additional components: 'addons-management/kapp-controller' ...
Verifying kubernetes version...
Retrieving configuration for upgrade cluster...
Create InfrastructureTemplate for upgrade...
Upgrading control plane nodes...
Patching KubeadmControlPlane with the kubernetes version v1.22.9+vmware.1...
Updating the KCP object with k8s version v1.22.9+vmware.1
Waiting for kubernetes version to be updated for control plane nodes
Upgrading worker nodes...
Patching MachineDeployment with the kubernetes version v1.22.9+vmware.1...
Waiting for kubernetes version to be updated for worker nodes...
unable to create AWS client. Skipping validations that require an AWS client
updating additional components: 'metadata/tkg' ...
unable to create AWS client. Skipping validations that require an AWS client
updating additional components: 'addons-management/standard-package-repo' ...
Waiting for packages to be up and running...
Cluster 'tkg-aws-workload-cluster' successfully upgraded to kubernetes version 'v1.22.9+vmware.1'
```

Figure 7.31 – Cluster upgrade log

The following screenshot shows the recycling of the workload cluster nodes to create the new version on the AWS EC2 console. Here, you can see that the old nodes got terminated and that the new nodes with newer versions were created to replace them:

Name	▲	Instance ID	Instance state	▽	Instance type	▽
tkg-aws-workload-cluster-control-plane-bwlrz		i-0d096a46a1c57f0ba	⊖ Terminated ⊕⊖		t3.large	
tkg-aws-workload-cluster-control-plane-v1-22-9-vmware-1-cvrvs2h		i-070d2f53f5d716c81	⊖ Terminated ⊕⊖		t3.large	
tkg-aws-workload-cluster-control-plane-v1-22-9-vmware-1-cvthzst		i-030b1f7960fa9f5b7	⊘ Running ⊕⊖		t3.large	
tkg-aws-workload-cluster-md-0-l5zl4		i-03b9d7f74d5e42ea5	⊖ Terminated ⊕⊖		t3.large	
tkg-aws-workload-cluster-md-0-nrmc8		i-092be694c504019bc	⊖ Terminated ⊕⊖		t3.large	
tkg-aws-workload-cluster-md-0-s7t8l		i-0655711379df414de	⊖ Terminated ⊕⊖		t3.large	
tkg-aws-workload-cluster-md-0-v1-22-9-vmware-1-pe5wg-bq45v		i-078d97bac3162ddc3	⊘ Running ⊕⊖		t3.large	
tkg-aws-workload-cluster-md-0-v1-22-9-vmware-1-pe5wg-d9fcr		i-02f7a631f5234a3f5	⊘ Running ⊕⊖		t3.large	
tkg-aws-workload-cluster-md-0-v1-22-9-vmware-1-pe5wg-h4jq6		i-0b1eac85b961f6cb3	⊘ Running ⊕⊖		t3.large	

Figure 7.32 – Cluster node recycling on the AWS EC2 console

5. Run the following command to ensure the workload cluster is running with the newer Kubernetes version:

```
$ tanzu cluster list
```

The following screenshot shows that the workload cluster is running on Kubernetes version 1.22.9:

```
~ » tanzu cluster list
  NAME                         NAMESPACE  STATUS   CONTROLPLANE  WORKERS  KUBERNETES       ROLES   PLAN
  tkg-aws-workload-cluster     default    running  1/1           3/3      v1.22.9+vmware.1  <none>  dev
```

Figure 7.33 – Confirming the workload cluster's upgrade

In addition to allowing the Kubernetes release to be upgraded, the `tanzu cluster upgrade` command also allows you to upgrade the cluster for a specific operating system and its versions. Run the `tanzu cluster upgrade -help` command to learn more about it. In addition to upgrading a TKG cluster for these reasons, there is also another dimension for the upgrades – upgrading the TKG version itself. Upgrading a TKG version (from TKG v1.5.x to v1.5.y or from v1.4.x to v1.5.y) is beyond the scope of this book. However, you can learn more about that here: `https://docs.vmware. com/en/VMware-Tanzu-Kubernetes-Grid/1.5/vmware-tanzu-kubernetes- grid-15/GUID-upgrade-tkg-index.html`.

Deleting a Tanzu Kubernetes Grid workload cluster

Destructions are always easier than constructions. This is the same case with TKG workload clusters. The following simple steps outline the procedure to delete the TKG workload cluster that we have used so far in this chapter:

1. Switch the `kubectl` context so that it's pointing to the management cluster that we previously created, which we used to create the workload cluster:

    ```
    $ kubectl config use-context tkg-aws-mgmt-cluster-admin@
    tkg-aws-mgmt-cluster
    ```

2. Run the following command to delete the workload cluster, along with its resources on your AWS account:

    ```
    $ tanzu cluster delete tkg-aws-workload-cluster
    ```

    You should see the following confirmation message on the console for the cluster deletion in progress:

```
~ » tanzu cluster delete tkg-aws-workload-cluster
Deleting workload cluster 'tkg-aws-workload-cluster'. Are you sure? [y/N]: y
Workload cluster 'tkg-aws-workload-cluster' is being deleted
```

Figure 7.34 – Cluster deletion in progress

The following screenshot from the AWS EC2 console shows that all the nodes for the workload clusters have been terminated:

Name	▲	Instance ID	Instance state	▽	Instance type	▽
tkg-aws-workload-cluster-control-plane-bwlrz		i-0d096a46a1c57f0ba	⊖ Terminated ⊕ ⊖		t3.large	
tkg-aws-workload-cluster-control-plane-v1-22-9-vmware-1-cvrvs2h		i-070d2f53f5d716c81	⊖ Terminated ⊕ ⊖		t3.large	
tkg-aws-workload-cluster-control-plane-v1-22-9-vmware-1-cvthzst		i-030b1f7960fa9f5b7	⊖ Terminated ⊕ ⊖		t3.large	
tkg-aws-workload-cluster-md-0-l5zl4		i-03b9d7f74d5e42ea5	⊖ Terminated ⊕ ⊖		t3.large	
tkg-aws-workload-cluster-md-0-nrmc8		i-092be694c504019bc	⊖ Terminated ⊕ ⊖		t3.large	
tkg-aws-workload-cluster-md-0-s7t8l		i-0655711379df414de	⊖ Terminated ⊕ ⊖		t3.large	
tkg-aws-workload-cluster-md-0-v1-22-9-vmware-1-pe5wg-bq45v		i-078d97bac3162ddc3	⊖ Terminated ⊕ ⊖		t3.large	
tkg-aws-workload-cluster-md-0-v1-22-9-vmware-1-pe5wg-d9fcr		i-02f7a631f5234a3f5	⊖ Terminated ⊕ ⊖		t3.large	
tkg-aws-workload-cluster-md-0-v1-22-9-vmware-1-pe5wg-h4jq6		i-0b1eac85b961f6cb3	⊖ Terminated ⊕ ⊖		t3.large	

Figure 7.35 – Terminated cluster nodes on the AWS EC2 console

Along with the EC2 instances, TKG (with the help of the Cluster API) also deletes other network resources that have been created for the deleted cluster on your AWS account. As you may have assumed, deleting a Kubernetes cluster is an extremely sensitive operation that could result in extended downtime for the applications running on it. Extensive measures should be taken to prevent access to such operations from environments other than a lab. Now, let's look at an even more destructive operation: deleting the entire TKG foundation, including the management cluster.

Deleting a Tanzu Kubernetes Grid foundation

You would rarely need to delete a TKG foundation from its roots except in such a lab environment. Nevertheless, we will cover this TKG life cycle activity in this chapter. To delete a TKG foundation, we just need to delete the management cluster that we created for the same. And like deleting a workload cluster, deleting a management cluster is also a simple but highly destructive process. The following steps outline the procedure for this:

1. Ensure you are pointing to the right Kubernetes cluster for the kubectl context by running the following command:

    ```
    $ kubectl config use-context tkg-aws-mgmt-cluster-admin@
    tkg-aws-mgmt-cluster
    ```

2. Run the following command to delete the management cluster, along with its resources on your AWS account:

    ```
    $ AWS_REGION=us-east-1 tanzu mc delete --verbose 5
    ```

    You may need to replace the region name in the command based on the deployed region of the management cluster. Upon command execution, you should see the following logs on the console for the cluster deletion in progress because of the --verbose option, followed by the logging detail level. This command takes the log verbose level from 0 to 9, with 9 being the most detailed log:

```
Ready after 16s
Installing providers to cleanup cluster...
Fetching providers
Installing cert-manager Version="v1.5.3"
Waiting for cert-manager to be available...
Installing Provider="cluster-api" Version="v1.0.1" TargetNamespace="capi-system"
Installing Provider="bootstrap-kubeadm" Version="v1.0.1" TargetNamespace="capi-kubeadm-bootstrap-system"
Installing Provider="control-plane-kubeadm" Version="v1.0.1" TargetNamespace="capi-kubeadm-control-plane-system"
Installing Provider="infrastructure-aws" Version="v1.2.0" TargetNamespace="capa-system"
installed   Component=="cluster-api"  Type=="CoreProvider"   Version=="v1.0.1"
installed   Component=="kubeadm"  Type=="BootstrapProvider"   Version=="v1.0.1"
installed   Component=="kubeadm"  Type=="ControlPlaneProvider"   Version=="v1.0.1"
installed   Component=="aws"  Type=="InfrastructureProvider"   Version=="v1.2.0"
Waiting for provider control-plane-kubeadm
Waiting for provider cluster-api
Waiting for provider bootstrap-kubeadm
Waiting for provider infrastructure-aws
Waiting for resource capi-kubeadm-bootstrap-controller-manager of type *v1.Deployment to be up and running
Waiting for resource capi-kubeadm-control-plane-controller-manager of type *v1.Deployment to be up and running
Waiting for resource capi-controller-manager of type *v1.Deployment to be up and running
Waiting for resource capa-controller-manager of type *v1.Deployment to be up and running
Passed waiting on provider bootstrap-kubeadm after 5.141313736s
Passed waiting on provider cluster-api after 5.179712151s
Passed waiting on provider infrastructure-aws after 5.265423202s
Passed waiting on provider control-plane-kubeadm after 10.14069577s
Success waiting on all providers.
Moving Cluster API objects from management cluster to cleanup cluster...
Performing move...
Discovering Cluster API objects
Moving Cluster API objects Clusters=1
Creating objects in the target cluster
Deleting objects from the source cluster
Waiting for the Cluster API objects to get ready after move...
Waiting for resource tkg-aws-mgmt-cluster of type *v1beta1.Cluster to be up and running
Waiting for resources type *v1beta1.MachineList to be up and running
Deleting management cluster...
Waiting for tkg-aws-mgmt-cluster resource of type *v1beta1.Cluster to be deleted
Management cluster 'tkg-aws-mgmt-cluster' deleted.
Deleting the management cluster context from the kubeconfig file '/Users/panditpa/.kube/config'
warning: this removed your active context, use "kubectl config use-context" to select a different one
Deleting kind cluster: tkg-kind-cbur8ctvqc7r4crv0i20

Management cluster deleted!
```

Figure 7.36 – Management cluster deletion logs

You may have figured out from the logs that TKG created a kind cluster on the bootstrap machine to do all the required cleanup to delete the TKG foundation from the AWS account. This is the same approach that TKG uses while creating a management cluster, as we saw earlier in this chapter.

With that, we have completed some of the important day-2 activities around TKG clusters. Now, let's wrap up this chapter with a quick summary of what we have learned.

Summary

At the beginning of this chapter, we discussed some of the reasons why TKG could be a good choice for being a Kubernetes-based container platform. As we saw during the hands-on activities, TKG makes Kubernetes platform deployment and management very easy and operationally efficient by providing a uniform interface – the Tanzu CLI. All the Tanzu CLI-based operations we performed in this chapter were infrastructure-agnostic, providing the required muti-cloud ease of operations.

Because of the limited scope of TKG in this book, we could not install and use all the optional extensions that TKG provides, but we covered them briefly to understand their applications. We saw how extensively TKG uses various cherry-picked open source tools from the CNCF ecosystem. This way, TKG is a solution completely backed by the open source community.

Finally, we learned about the common day-1 and day-2 activities on the TKG platform, starting with installing a platform on AWS and creating a workload cluster to host actual application containers. Following this, we learned how to add more capacity for that workload cluster with on-demand scaling. We also learned how easily we can upgrade the cluster for different reasons and finally how to delete the cluster and the foundation if required.

As you may have assumed, we have not covered many topics around this subject to keep this chapter's length concise. However, we will learn about TKG clusters' backup and restore, compliance scanning, and governance policy configurations in *Chapter 9, Managing and Controlling Kubernetes Clusters with Tanzu Mission Control,* for Tanzu Mission Control, a single pane of glass that controls hundreds of Kubernetes clusters.

TKG is a commercially available licensed software provided by VMware. In the next chapter we will go deep into the Tanzu developer experience with **Tanzu Application Platform**.

8

Enhancing Developer Productivity with Tanzu Application Platform

In the world of enterprise software, we have the concept of *Big A* applications and *Little A* applications. *Big A* might be a giant corporate billing system with hundreds of components, whereas a *Little A* application might be a single job that pulls records from a mainframe and writes them to MongoDB.

Similarly, there are *Big P* and *Little P* problems that need to be solved when delivering software in the enterprise. An example of a *Little P* problem I faced today was moving some container images from the VMware corporate container registry to a customer's private registry so they could use them internally without having to allow egress to the internet. On the other hand, some *Big P* problems you might face in the enterprise space might be the following:

- Making a company's developers measurably more productive

- Getting applications into production quickly, safely, and consistently – minimizing toil and blockers

- Getting developers, operators, and security specialists all pointed in the same direction and focused on the same problems

- Delivering applications to modern container platforms such as Kubernetes when developers have limited exposure and experience

Coincidentally, those are exactly the four problems we talk about when we discuss **Tanzu Application Platform**.

In this chapter, we will cover these topics:

- Why should I use Tanzu Application Platform?

- The building blocks of Tanzu Application Platform

- Day 1 – installing and configuring Tanzu Application Platform
- Day 2 – deploying applications to Tanzu Application Platform
- Day 3 – common operational activities on Tanzu Application Platform
- Next steps

Now that we know what we're going to accomplish in this chapter, let's jump in and talk about the reasons you and your team might need to use Tanzu Application Platform.

Why should I use Tanzu Application Platform?

As I mentioned in the chapter introduction, there are small problems that can be encountered when writing software, such as efficiently sorting a list in place or moving streaming data from a legacy database to the latest NoSQL offering, and there are big problems, such as the following:

- **Developer productivity**: Enterprise software only exists because big companies hire software developers to write the software that they use to bring in revenue and differentiate the company from their competition. Companies with good platforms can focus their developers' time and effort on meaningful tasks that directly affect the company's bottom line. This makes the developers feel valued. Those without a good platform grow to view their developers as an expensive cost center, sparking a painful downward spiral into low morale and low productivity.

- **Getting software into production**: If you haven't written software for a large company, you may be surprised at how difficult it is to get a piece of software into production. There are often dozens of manual, tedious tasks to check all the boxes for security, compliance, and downstream risk. The enlightened players in this space have figured out how to automate away the tedium, building these steps into an automated process, abstracted away from developers.

- **Align developers, operators, and security toward the same goal**: By building compliance, governance, best practices, and security into a streamlined automated process, you eliminate the friction often found between developers, operators, and security specialists. While the operators push back against any sort of change, security specialists want releases scrutinized down to the last bit. This can swallow up weeks of time and effort on a development team while crushing morale. By building checks and controls into the software supply chain, operators and security specialists can rest easy that their guidance is being followed, while developers are freed up to deliver software at a consistently high velocity.

- **On-ramp to Kubernetes**: Enterprise adoption of Kubernetes is on an upswing and shows few signs of slowing down. Kubernetes is a very powerful platform that enables some big outcomes. However, it comes with considerable complexity and a steep learning curve. Tanzu Application Platform abstracts away some of this complexity, allowing software teams to leverage Kubernetes while benefitting from some abstractions that simplify the path to production.

To summarize, Tanzu Application Platform is an end-to-end supply chain for delivering software to production safely, securely, and reliably at scale. Now that we know at a high level what it is, let's proceed to break it down and look at the individual components.

The building blocks of Tanzu Application Platform

Tanzu Application Platform is an opinionated set of technologies working together to deliver significant outcomes for platform operators, developers, and security professionals. Covering each component in depth would require more space than this chapter allows, so I encourage you to visit the official documentation for more detail on any of these components.

The diagram that follows offers a high-level view of Tanzu Application Platform:

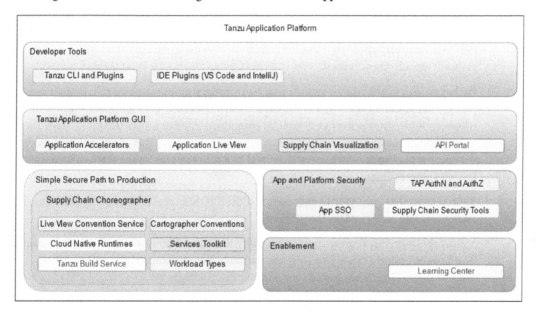

Figure 8.1 – Tanzu Application Platform components

This diagram attempts to group and categorize the various components. Here's a breakdown of the products that follow the hierarchy from the diagram:

- **Developer Tools**: In the previous section, we mentioned that Tanzu Application Platform is a ramp for developers to get onto Kubernetes, as well as a platform designed to make them as efficient and productive as possible. Here are the components that help accomplish this:

  - **Tanzu CLI**: Unless this is the first chapter of this book that you're reading, you'll have seen this tool before. The Tanzu CLI is a tool that allows developers to deal with higher-level abstractions rather than interact directly with infrastructure and Kubernetes **Custom Resources (CRs)**. It does so via *plugins* specific to a particular feature.

- **Apps CLI plugin**: This plugin to the Tanzu CLI deals with two abstractions that will become very familiar to developers on the platform: *cluster supply chains* and *workloads*. A *supply chain* is a Kubernetes CR representing all the steps from the developer's desktop to production application. A *workload* is the application itself that makes the journey into production via a supply chain. This CLI plugin lets you get a workload into production without manually configuring YAML files.

- **Insight CLI plugin**: As we mentioned in the previous point, you can use the Tanzu CLI to deploy an application and run it through a supply chain, which will move it onto the platform as a running app. Many supply chains include steps that scan resources for vulnerabilities. This plugin allows you to manually run those scans or report on the results of those scans. The four entities that this plugin allows you to work with are as follows:

  - **Images**: This is a container image that exists in a registry. Often, these images are built as part of a supply chain, but they don't have to be. You can send a **Software Bill Of Materials (SBOM)** for scanning or querying the results of previous scans against an image that the platform already knows about.

  - **Packages**: This is just what it sounds like. Many programming languages have their own package management and distribution mechanisms. The same goes for operating systems, whose root filesystems may very well be embedded in a container image. Most of the vulnerabilities that get surfaced in a scan are in these packages. This plugin lets you query the specific packages, which were identified in a source or an image scan, that contain known vulnerabilities.

  - **Source**: This allows you to scan your source code's SBOM in real time or query past source code scans that happened on the platform.

  - **Vulnerabilities**: This represents a vulnerability that turned up in a source or image scan. The Insight plugin allows you to query individual vulnerabilities (e.g., Log4Shell or Log4Spring) and tie them back to specific scans or packages.

- **Developer IDE plugins**: Tanzu Application Platform provides IDE plugins for both **Visual Studio Code (VS Code)** and *IntelliJ IDEA*. As developers move toward a distributed cloud-native model, it becomes increasingly difficult to replicate the runtime environment on their desktops. Tanzu Application Platform provides the ability to quickly iterate on and debug their code on a running Kubernetes cluster. Here is a current list of those IDE plugins:

  - **Direct Deploy**: This saves developers the steps of piecing together a complex `tanzu` CLI command or a `workload.yaml` file. Developers can deploy workloads directly to a Kubernetes cluster from their IDEs.

  - **Live Update**: Developers can make a local change on their workstations and see it running live on a Kubernetes cluster in seconds. This uses *Tilt* under the covers to sync the local code base with the workload running remotely on Kubernetes.

- **Remote Debug**: Sometimes log messages just aren't sufficient and you need to step through your code, examining local and wider-scoped variables in the process. As we just mentioned, this is straightforward on a local workstation, but in a modern environment, this often isn't sufficient to debug an issue occurring in Kubernetes. Directly debugging a workload on Kubernetes often entails a messy array of startup parameters and port forwards. This plugin automates the whole process, giving developers a one-click solution to remote debugging.

- **Running Workloads**: With this VS Code-only feature, developers get a panel inside their IDE to visualize all workloads running in their current Kubernetes context.

- **Tanzu Application Platform GUI**: The Tanzu Application GUI, based on the *Backstage* open source project, is a central dashboard for Tanzu Application Platform. It provides some out-of-the-box functionality that it gets from Backstage, as well as a number of very useful custom plugins that are only available to Tanzu Application Platform users. Here's a quick peek at what you can expect to see in the Tanzu Application Platform GUI, starting with a visual overview of the UI and its layout:

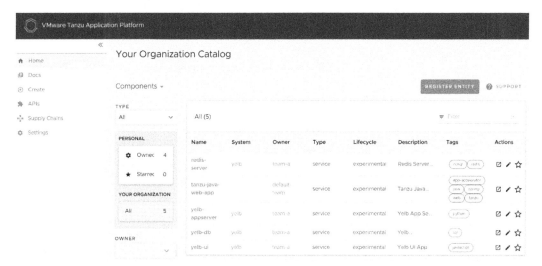

Figure 8.2 – Tanzu Application Platform GUI

This is the screen you'll see when first accessing the Tanzu Application GUI. The following is a breakdown of the GUI's components:

- **Organization Catalog**: Tanzu Application Platform GUI provides a central repository for an entire organization to publish and catalog their software. If we dig into **tanzu-java-web-app** in the preceding screenshot, we'll see the following screen:

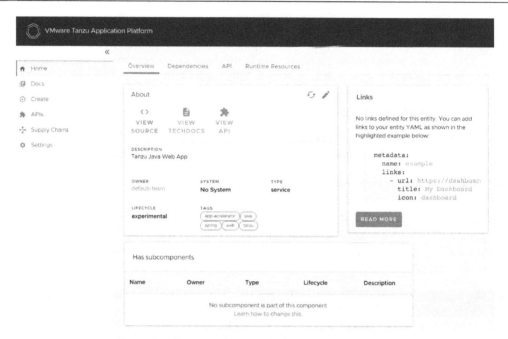

Figure 8.3 – Tanzu Application Platform GUI catalog entry

The catalog entry allows teams to provide a "one-stop shop" for their application's source code, tech docs, API definition, and any relevant links:

- **TechDocs**: This is a backstage plugin that allows development teams to create documentation in Markdown format, and will present that documentation as part of the application's entry in the organization catalog.

- **Runtime Resources Visibility**: Users of other Tanzu products, such as *Tanzu Application Service* (`https://tanzu.vmware.com/application-service`), have benefitted from being able to see all their runtime resources in one place. The Tanzu team brought this to Tanzu Application Platform GUI as well. Anything associated with a registered catalog entity will be visible here.

- **Application Live View**: This feature works with other components, such as *Convention Service*, to provide deep insight into the runtime state of a running application.

- **Application Accelerator**: This feature, which we covered in depth in *Chapter 2, Developing Cloud-Native Applications*, is part of Tanzu Application Platform GUI.

- **API Documentation**: This feature, which we covered in depth in *Chapter 5, Defining and Managing Business APIs*, is also part of Tanzu Application Platform GUI.

- **Supply Chain Choreographer Visualization**: Supply Chain Choreographer, which lies at the heart of Tanzu Application Platform, is just what it sounds like, a chained set of inputs and outputs that guide an application from its source to a running application in a production environment. This plugin provides a compelling visualization of those supply chains.

- **Supply Chain Choreographer**: If you think of Tanzu Application Platform GUI as the "eyes and ears" of the platform, then Supply Chain Choreographer would be its beating heart. Supply Chain Choreographer is based on the open source *Cartographer* project (`https://github.com/vmware-tanzu/cartographer`), and it allows platform operators and software architects to preconfigure and pre-approve multiple paths to production. There might be a path to production for Spring Boot APIs that fall under PCI controls and another path to production for event-driven data processing applications. Just as was the case with Tanzu Application Platform GUI, Supply Chain Choreographer is a feature of the platform that bundles and contains other features. Here's a list of some of the platform features that fall within the realm of Supply Chain Choreographer:

  - **Live View Convention Service**: One of the features mentioned in our discussion of Tanzu Application Platform GUI was the *Application Live View* GUI plugin. For this plugin to work, an application needs to expose its inner workings via an opinionated, well-known API. A naïve approach to this might be to require all application teams to implement this themselves, by bringing in a particular version of Spring Boot Actuator with its own web listener running on a specific port with specific endpoint naming conventions. This would result in added burden on the development teams as well as the potential for misconfiguration. A better approach would be to modify every workload as it passes through the supply chain to expose the inner workings in a consistent, opinionated way. That's exactly what this service does. It sits in the supply chain and adds consistent Java or .NET Core configuration parameters to every eligible application that uses the supply chain.

  - **Cartographer Conventions**: Application Live View isn't the only feature for which supply chain conventions are a good fit. Any features that a platform operator or a DevOps lead wants to apply across the entire portfolio are a good fit here. One example is labeling all application artifacts with the owner of the project. Another might be to configure sane upper and lower limits to an app that auto-scales with *Cloud Native Runtimes*. The possibilities are endless, and Cartographer Conventions is a convenient way to apply conventions across a wide range of applications running on the platform.

  - **Cloud Native Runtimes**: This is the component that runs your deployable application artifacts on Kubernetes. It is based on the open source *Knative* project (`https://github.com/knative`), which, while it is often thought of as a serverless runtime, is also an excellent way to run regular web-facing workloads. It has a unique approach to load balancing that allows for auto-scaling based on configurable metrics while minimizing lost or dropped requests.

  - **Services Toolkit**: This is another feature that evolved from the much-loved *Tanzu Application Service*. That product allowed users to request a service to be provisioned, perhaps a database or a message queue, and bind the credentials of that service to an application at runtime. *Services Toolkit* brings that same functionality to Kubernetes. You can think of it as data services for developers who, first, don't want to think about managing data services, and second, don't want to manually wire up their application to those data services.

- **Tanzu Build Service**: In a previous bullet, we talked about running deployable artifacts. Tanzu Build Service is the component that takes an application's source code as input and outputs a deployable artifact in the form of an OCI container image. We did a thorough deep-dive into this feature in *Chapter 3, Building Secure Container Images with Build Service*.

- **Workload Types**: You can think of this as a set of preconfigured, pre-curated supply chains that you can use to get started quickly with several different kinds of workloads:

  - **Web**: This is a standard web-facing application that you want to be able to scale up and down

  - **TCP**: This is a good fit for running a legacy application that handles its own web interactions and you want to pass network traffic directly to it

  - **Queues**: These are applications that run in the background and process events as they arrive on a queue

  - **Functions**: These are a hybrid of web and queue applications that allow developers to implement a single piece of functionality that only gets instantiated and called when a request comes in

- **App and Platform Security**: Here are some of the components that make security a first-class citizen on Tanzu Application Platform:

  - **Tanzu Application Platform Authentication and RBAC Authorization**: Tanzu Application Platform GUI uses the single sign-on functionality that comes with Backstage. This allows platform operators to configure providers such as Okta, Google, Azure, and GitHub. Furthermore, the GUI can be configured such that it can monitor running resources across multiple Kubernetes clusters, not just the cluster it is running on.

  - **App single sign-on**: This is a very common request among application developers. They don't want to manage their own login, user **authentication** (**authN**), or **authorization** (**authZ**). Rather, they want to delegate those functions to the platform and have it taken care of for them. That's exactly what this feature does. It allows platform operators to stand up preconfigured, opinionated deployments of Spring Authorization Server, and make that server available to applications via integrations into the platform's software supply chains. The idea of this service is to integrate single sign-on into the application from the very beginning rather than as a last-minute bolt-on solution.

  - **Supply Chain Security Tools**: Software supply chain attacks are top-of-mind for many in the enterprise software space, and for good reason. Supply Chain Security Tools is a suite of tools that plug directly into a software supply chain such that whenever any part of that supply chain changes, the source code, bundled dependencies, and generated container images get scanned for all known vulnerabilities using a constantly updated database. These tools handle scanning the artifacts, storing the scan results, reporting on the stored results, and cryptographically signing the generated artifacts so they can't be changed after they've been scanned. This allows platform operators to enable best-of-breed security for all their running workloads simply by incorporating them into a supply chain.

- **Enablement**: Any product of the size and scope of Tanzu Application Platform needs a way for users and operators of the product to get up to speed. Tanzu Application Platform bundles its own learning platform for enabling those users. This allows platform operators to quickly and easily set up hands-on workshops that walk the various personas through the product. It also includes a brief built-in workshop on how to build your own workshops. Here's a quick look at **Learning Center for Tanzu Application Platform** showing my favorite feature, the browser-embedded VS Code editor:

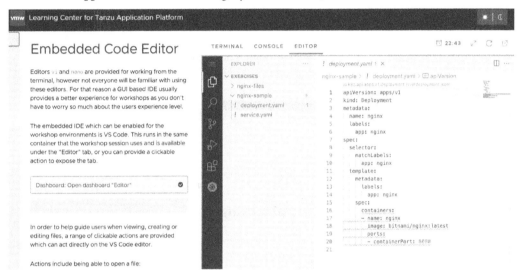

Figure 8.4 – Learning Center with the Embedded VS Code editor

This allows architects and operators to build hands-on real-world workshops that automatically provision and run completely self-contained in the browser. Here's a quick list of some of the available features:

- **Fully functional console**: Any workshop tasks that must happen in a shell in a console can be executed from within the browser. Furthermore, you can provide a link that will paste code directly into the console and execute it. Furthermore, these consoles can be preconfigured with `kubectl` pointing to a live Kubernetes cluster and embedded *Docker*, so anything requiring the Docker CLI, builds, runs, pushes, and so on, are also possible.

- **Embedded comprehension questions**: Workshop builders can embed questions right into the workshop to make sure users are properly understanding the concepts.

- **Embedded VS Code editor**: Not only can workshop builders embed VS Code right into their workshop, but they can provide links that will open files, select code blocks in the editor window, and even paste additional fragments in just the right location.

- **Embedded web pages**: This allows users to deploy something to their dedicated workshop Kubernetes instance and immediately see it running right in the same window. It also enables tools like the Kubernetes web console to be exposed.

- **Verify Workshop State**: Users can click a link that will trigger Kubernetes commands that don't appear in a console. This allows workshop developers to ensure that the workshop state is as desired before moving on.

- **Docker builds**: The embedded console uses **Docker in Docker** (**DIND**) to allow for the full range of Docker commands.

- **Download files**: Workshop developers can enable users to download a local copy of any or all workshop files for future reference.

Now that we've covered whom Tanzu Application Platform is targeted to and what it consists of, it's time to get our hands dirty and install it, which is exactly what we'll do in the next section.

Day 1 – installing and configuring Tanzu Application Platform

To install Tanzu Application Platform, we'll need a few things in place first.

Prerequisites and technical requirements

In order to install and run Tanzu Application Platform, there are some technical prerequisites that must be in place. Here's a list of what is required:

- **A (large) Kubernetes cluster**

  The Tanzu Application Platform documentation lists several supported Kubernetes clusters, including Tanzu Kubernetes Grid and Minikube. However, to keep these instructions manageable, I'm going to recommend three public cloud Kubernetes clusters. These clusters are as follows:

  - **Amazon Elastic Kubernetes Service (EKS)**

  - **Azure Kubernetes Service (AKS)**

  - **Google Kubernetes Engine (GKE)**

  There are a number of reasons for this. Here are a few:

  - Easy out-of-the-box load balancer services. While we can make Tanzu Application Platform work with node ports, load balancer services make things much easier.

  - Easy out-of-the-box persistent volume claims. The same reasoning behind easy load balancers also applies to dynamic storage, which is also required for Tanzu Application Platform.

  - Easy (albeit pricey) access to sufficient resources. While you can install Tanzu Application Platform on Minikube on a laptop, you'll come up against RAM and CPU limitations. The public cloud options make it easy to stand up a cluster that can scale up to a size that runs Tanzu Application Platform without issue.

The Tanzu Application Platform documentation has some recommendations for RAM and CPU, but here's what has worked for me:

- One or more control plane nodes. If your Kubernetes cluster gives you access to the control plane, the node should have at least two vCPUs and 8 GB of RAM.

- Five or more worker nodes. Each node should have at least two vCPUs and 8 GB of RAM.

- **A Tanzu Network account**

The software installation is via container images hosted on the container registry associated with the Tanzu Network. You can sign up here: `https://network.tanzu.vmware.com/`.

- **A production container registry**

Once again, there are multiple options here, but I'll recommend two that will help things go more smoothly. These require paid accounts, but they greatly simplify the installation. I'll point out that Amazon Elastic Container Registry isn't yet supported. Here are the currently supported options for a container registry:

- **A paid Docker Hub account**

Docker Hub does offer a free tier, but the Tanzu Application Platform install requires a significant amount of space (at least 1 GB) and significant repeated image pushes and pulls. Docker limits both on their free tier and this will hinder smooth operation.

- **Google Container Registry (gcr.io)**

This is my preferred solution as you can quickly set up a registry for your GCP project and push to multiple repositories.

- **Azure Container Registry (azurecr.io)**

This is another good option, especially if you have an Azure account already set up.

- **Harbor (DIY)**

I'd urge caution in using bring-your-own Harbor unless you're comfortable standing up trusted TLS with a public **certificate authority (CA)** or *Let's Encrypt*. If your Harbor presents a self-signed certificate, the Tanzu Application Platform installation gets more complicated. Harbor works great as a container registry, but if you don't already have it up and running, you'll need to spend some time getting it properly set up.

- **A custom domain with a DNS server**

All the public cloud providers will register a domain for you and provide DNS lookup for that domain. You can also do this with dedicated registrars such as GoDaddy. The only requirement here is that you're able to create wildcard *A* and *CNAME* records that resolve publicly.

- **(Optional but recommended) A production Git server**

  You can get through the demos without your own Git server or an account on a public service, but any significant real-world work with Tanzu Application Platform requires one. There are free options on GitLab (gitlab.com) and GitHub (github.com) that work great for our purposes. You simply need to create an account and get a set of SSH or user/password credentials that the platform can use to push to and pull from a remote Git repo.

- The kubcectl CLI as well as the *Carvel* tools. You can find them at https://kubernetes. io/docs/tasks/tools/ and https://carvel.dev/#install, respectively. Kubectl should be version 1.22 or 1.23. It should also be authorized with cluster-admin privileges on the Kubernetes cluster you'll be installing to. All the Carvel tools are useful, but for this install, you only need imgpkg. You'll also need the Docker CLI installed such that you don't need to be root to run it.

Let's do some quick spot-checks that we have all our prerequisites in place:

- Let's double-check our *Docker* for the dreaded **permission denied** error. This is what you can expect to see if your Docker daemon doesn't allow non-root access:

  ```
  rob@rob-virtual-machine2:~$ docker info # no good, notice
  'permission denied'
  Client:
   Context:    default
   Debug Mode: false


  Server:
  ERROR: Got permission denied while trying to connect to
  the Docker daemon socket at unix:///var/run/docker.sock:
  Get http://%2Fvar%2Frun%2Fdocker.sock/v1.24/info: dial
  unix /var/run/docker.sock: connect: permission denied
  errors pretty printing info
  version.BuildInfo{Version:"v3.5.3",
  GitCommit:"041ce5a2c17a58be0fcd5f5e16fb3e7e95fea622",
  GitTreeState:"dirty", GoVersion:"go1.15.8"}
  ```

  This is what a successful installation looks like:

  ```
  rob@rob-virtual-machine2:~$ docker info # all good!
  Client:
   Context:    default
   Debug Mode: false
  ```

```
Server:
 Containers: 0
  Running: 0
```

- Finally, let's verify `kubectl` and our Carvel `imgpkg` tool:

```
ubuntu@ip-172-31-33-59 ~> kubectl version
Client Version: version.Info{Major:"1",
Minor:"22", GitVersion:"v1.22.13",
GitCommit:"a43c0904d0de10f92aa3956c74489c45e6453d6e",
GitTreeState:"clean", BuildDate:"2022-08-17T18:28:56Z",
GoVersion:"go1.16.15", Compiler:"gc", Platform:"linux/
amd64"}
Server Version: version.Info{Major:"1",
Minor:"22+", GitVersion:"v1.22.11-eks-18ef993",
GitCommit:"b9628d6d3867ffd84c704af0befd31c7451cdc37",
GitTreeState:"clean", BuildDate:"2022-07-06T18:06:23Z",
GoVersion:"go1.16.15", Compiler:"gc", Platform:"linux/
amd64"}
ubuntu@ip-172-31-33-59 ~> imgpkg version
imgpkg version 0.27.0
```

Once your command-line tools are in order, you're ready to proceed with installation in the next section.

Accepting end user license agreements

Before we can commence with moving packages to our custom registry and installing them, we need to accept the **End User License Agreements** (**EULAs**). After logging into the Tanzu Network, navigate to the two products listed here and select the latest version of each. If you need to accept the EULA, you will see a yellow box right below the version dropdown. You can click the link in the yellow box to be prompted to accept the EULA:

- `https://network.tanzu.vmware.com/products/tanzu-application-platform/`

- `https://network.tanzu.vmware.com/products/tanzu-cluster-essentials/`

Relocating Tanzu Application Platform packages

While this step isn't absolutely necessary, it's highly recommended to move all of the container images from the Tanzu repository to your own. If you think about how Kubernetes works, when any of the dozens (or hundreds!) of pods that constitute Tanzu Application Platform needs to start or restart, the `kubelet` process on a worker node will need to pull down that pod's container images. If you don't relocate those container images to your own registry, they'll have to come directly from the Tanzu Network container registry, meaning you'll have `registry.tanzu.vmware.com` in your runtime critical path. That means anytime you need to restart a pod on a Kubernetes worker that doesn't already contain that pod's container images, the Tanzu registry must be up and responsive. Now, the Tanzu team has very good SREs who do an excellent job of keeping the Tanzu registry up and responsive, but their **Service Level Objectives (SLOs)** may not line up with your apps' SLOs, so it's a good idea to manage your own destiny and maintain your own container registry. Furthermore, the Tanzu registry is shared with every other Tanzu customer, so you can't rule out spikes in demand making container images temporarily unresponsive.

We'll use the `imgpkg` tool to copy all the Tanzu Application Platform images from the Tanzu registry into our custom registry. The examples here will use `gcr.io`, but the steps are very similar for Azure Container Registry, Harbor, or Docker Hub.

First, let's use the Docker CLI to log in to the Tanzu registry:

```
ubuntu@ip-172-31-33-59 ~> docker login registry.tanzu.vmware.
com
Authenticating with existing credentials...
Login Succeeded
ubuntu@ip-172-31-33-59 ~>
```

Then, log in once again to your personal registry. Here's how you do it for `gcr.io`:

```
ubuntu@ip-172-31-33-59 ~/t/packt> cat ./gcr-json-key.json |
docker login -u _json_key --password-stdin gcr.io
Login Succeeded
```

For Docker Hub, Harbor, or Azure, you can issue `docker login <registry-host>` and type a username and password in at the prompts. At this point, we are authenticated to both the Tanzu registry and our custom registry (`gcr.io`). The `imgpkg` tool requires the local Docker daemon to be authenticated before it can copy over the images.

Now, we can use `imgpkg` to pull images from the Tanzu registry and push them to our custom registry. We'll set some environment variables for readability:

```
ubuntu@ip-172-31-33-59:~/tap/packt$ export INSTALL_REGISTRY_
USERNAME=_json_key
ubuntu@ip-172-31-33-59:~/tap/packt$ export INSTALL_REGISTRY_
PASSWORD=$(cat gcr-json-key.json)
ubuntu@ip-172-31-33-59:~/tap/packt$ export INSTALL_REGISTRY_
HOSTNAME=gcr.io
```

```
ubuntu@ip-172-31-33-59:~/tap/packt$ export INSTALL_REGISTRY_
PROJECT_NAME=my-gcr-project
ubuntu@ip-172-31-33-59:~/tap/packt$ export TAP_VERSION=1.3.2
ubuntu@ip-172-31-33-59:~/tap/packt$ imgpkg copy -b registry.
tanzu.vmware.com/tanzu-application-platform/tap-packages:${TAP_
VERSION} --to-repo ${INSTALL_REGISTRY_HOSTNAME}/${INSTALL_
REGISTRY_PROJECT_NAME}/tap-packages
copy | exporting 177 images...
copy | will export registry.tanzu.vmware.com/tanzu-applica-
tion-platform/tap-packages@sha256:001224ba2c37663a9a412994f-
1086ddbbe40aaf959b30c465ad06c2a563f2b9f
…many, many more lines
copy | exported 177 images
copy | importing 177 images...


 6.77 GiB / 6.77 GiB [=========================================
=================================================================
====] 100.00% 194.00 MiB/s 35s


copy | done uploading images
Succeeded
```

At this point, all the container images needed to install and run Tanzu Application Platform are available at gcr.io/my-gcr-project/tap-packages. Specific instructions for logging in to and relocating images into Dockerub, Harbor, and Azure are available in the Tanzu Application Platform install documentation here: https://docs.vmware.com/en/VMware-Tanzu-Application-Platform/1.3/tap/GUID-install.html#relocate-images-to-a-registry-0.

Installing the Tanzu CLI and Tanzu Application Platform plugins

If you don't already have the Tanzu CLI installed from other chapters, you can do that now. Furthermore, even if you have the Tanzu CLI, we'll install some plugins to help us work more efficiently with Tanzu Application Platform.

First, we'll visit the Tanzu Application Platform page on Tanzu Network and download the CLI. The link is here: https://network.tanzu.vmware.com/products/tanzu-application-platform/#/releases/1222090/file_groups/10484.

Once you've downloaded the tanzu-framework-bundle tarfile for your platform, copy it to your home directory. Then, on Mac and Linux, you can follow these steps. On Windows, you may have to adapt them a bit, or, as I would recommend, use *VirtualBox* to stand up a Linux VM and follow the Linux instructions.

Here's the whole process. I'll add some comments inline to describe what's going on:

```
ubuntu@ip-172-31-33-59:~/downloads$ mkdir -p $HOME/tanzu
ubuntu@ip-172-31-33-59:~/downloads$ tar -xvf tanzu-framework-
linux-amd64.tar -C $HOME/tanzu # extract the tar file into the
working directory.
cli/
cli/core/
cli/core/v0.11.6/
cli/core/v0.11.6/tanzu-core-linux_amd64
cli/core/plugin.yaml
cli/distribution/
cli/distribution/linux/
cli/distribution/linux/amd64/
cli/distribution/linux/amd64/cli/
cli/distribution/linux/amd64/cli/accelerator/
cli/distribution/linux/amd64/cli/accelerator/v1.2.0/
cli/distribution/linux/amd64/cli/accelerator/v1.2.0/tanzu-
accelerator-linux_amd64
cli/distribution/linux/amd64/cli/package/
cli/distribution/linux/amd64/cli/package/v0.11.6/
cli/distribution/linux/amd64/cli/package/v0.11.6/tanzu-package-
linux_amd64
cli/distribution/linux/amd64/cli/apps/
cli/distribution/linux/amd64/cli/apps/v0.7.0/
cli/distribution/linux/amd64/cli/apps/v0.7.0/tanzu-apps-linux_
amd64
cli/distribution/linux/amd64/cli/secret/
cli/distribution/linux/amd64/cli/secret/v0.11.6/
cli/distribution/linux/amd64/cli/secret/v0.11.6/tanzu-secret-
linux_amd64
cli/distribution/linux/amd64/cli/insight/
cli/distribution/linux/amd64/cli/insight/v1.2.2/
cli/distribution/linux/amd64/cli/insight/v1.2.2/tanzu-insight-
linux_amd64
cli/distribution/linux/amd64/cli/services/
cli/distribution/linux/amd64/cli/services/v0.3.0/
cli/distribution/linux/amd64/cli/services/v0.3.0/tanzu-
```

```
services-linux_amd64
cli/discovery/
cli/discovery/standalone/
cli/discovery/standalone/apps.yaml
cli/discovery/standalone/services.yaml
cli/discovery/standalone/secret.yaml
cli/discovery/standalone/insight.yaml
cli/discovery/standalone/package.yaml
cli/discovery/standalone/accelerator.yaml
ubuntu@ip-172-31-33-59:~/downloads$ cd $HOME/tanzu
ubuntu@ip-172-31-33-59:~/tanzu$ export VERSION=v0.11.6
ubuntu@ip-172-31-33-59:~/tanzu$ sudo install cli/core/$VERSION/
tanzu-core-linux_amd64 /usr/local/bin/tanzu  # move the
executable into a well-known directory and make it executable.
ubuntu@ip-172-31-33-59:~/tanzu$ tanzu version
version: v0.11.6
ubuntu@ip-172-31-33-59:~/tanzu$ export TANZU_CLI_NO_INIT=true
# instruct the CLI not to look in external repositories for
plugins
ubuntu@ip-172-31-33-59:~/tanzu$ tanzu plugin install --local
cli all  # install the plugins from the distributable tar file
we just unpacked
Installing plugin 'accelerator:v1.2.0'
Installing plugin 'apps:v0.7.0'
Installing plugin 'insight:v1.2.2'
Installing plugin 'package:v0.11.6'
Installing plugin 'secret:v0.11.6'
Installing plugin 'services:v0.3.0'
∂  successfully installed 'all' plugin
ubuntu@ip-172-31-33-59:~/tanzu$ tanzu plugin list
  NAME                  DESCRIP-
TION                                            SCOPE       DISCOV-
ERY   VERSION  STATUS
  login                 Login to the plat-
form                                            Stan-
dalone  default    v0.11.6  update available
  management-cluster  Kubernetes management-cluster opera-
tions                                           Stan-
```

```
dalone   default    v0.11.6   update available
  package             Tanzu package manage-
ment                                              Stan-
dalone   default    v0.11.6   installed
  pinniped-auth       Pinniped authenti-
cation operations (usually not directly
invoked)                                  Stan-
dalone   default    v0.11.6   update available
  secret              Tanzu secret manage-
ment                                              Stan-
dalone   default    v0.11.6   installed
  services            Explore Service Instance Classes,
discover claimable Service Instances and manage Resource
Claims   Standalone           v0.3.0    installed
  apps                Applications on Kuber-
netes                                        Stan-
dalone              v0.7.0    installed
  insight             post & query image, package, source, and
vulnerability data                                Stan-
dalone              v1.2.2    installed
  accelerator         Manage accelerators in a Kubernetes
cluster                                    Stan-
dalone              v1.2.0    installed
```

Congratulations! If you made it this far, you have a working Tanzu CLI with all the plugins you'll need to install and run Tanzu Application Platform.

Installing Cluster Essentials

The Cluster Essentials toolset is how we land Tanzu packages on our Kubernetes cluster. Perhaps you're familiar with tools such as Helm. The Cluster Essentials tools do something similar, but in addition to what Helm does, they do the following:

- Store a desired application state in etcd as a CR so the application can be continuously reconciled

- Manage a complex hierarchy of dependencies

- Store and maintain templatized app configuration in such a way as to avoid configuration drift

- Allow selective sharing of secrets across namespaces

I won't sugarcoat the fact that Tanzu Application Platform is a complex piece of software, and it needs a fully realized enterprise-ready toolset to install it onto Kubernetes and keep it running smoothly. Cluster Essentials is that toolset.

To install Cluster Essentials, we follow a similar script to what we did for the CLI. In this case, we need to make sure that `kubectl` is pointing to the Kubernetes cluster we plan on deploying to as the install script will initiate the deployment. Once again, I'll narrate the process with comments:

```
ubuntu@ip-172-31-33-59:~/downloads$ mkdir -p $HOME/tanzu-clus-
ter-essentials  # create working directory
ubuntu@ip-172-31-33-59:~/downloads$ ls # check our downloaded
file
tanzu-cluster-essentials-linux-amd64-1.2.0.tgz
ubuntu@ip-172-31-33-59:~/downloads$ tar xvf ./tanzu-cluster-
essentials-linux-amd64-1.2.0.tgz -C $HOME/tanzu-cluster-essen-
tials    # unzip the downloaded file into the working directory
install.sh
uninstall.sh
imgpkg
kbld
kapp
ytt
ubuntu@ip-172-31-33-59:~/downloads$ cd $HOME/tanzu-cluster-es-
sentials
ubuntu@ip-172-31-33-59:~/tanzu-cluster-essentials$ kubectl
config get-contexts  # check our Kubernetes cluster
*        user@tap-install.us-west-2.eksctl.
io           tap-install.us-west-2.eksctl.
io                         user@tap-install.us-west-2.eksctl.
io
ubuntu@ip-172-31-33-59:~/tanzu-cluster-essentials$ export
INSTALL_BUNDLE=registry.tanzu.vmware.com/tanzu-cluster-essen-
tials/cluster-essentials-bundle@sha256:e00f33b92d418f49b1af79f-
42cb13d6765f1c8c731f4528dfff8343af042dc3e   # Set some env vars
for the install script to use
ubuntu@ip-172-31-33-59:~/tanzu-cluster-essentials$ export
INSTALL_REGISTRY_HOSTNAME=registry.tanzu.vmware.com
ubuntu@ip-172-31-33-59:~/tanzu-cluster-essentials$ export
INSTALL_REGISTRY_USERNAME=myuser@myemail.com
ubuntu@ip-172-31-33-59:~/tanzu-cluster-essentials$ export
INSTALL_REGISTRY_PASSWORD=<my_super_secret_password>
ubuntu@ip-172-31-33-59:~/tanzu-cluster-essentials$ ./install.
sh # this will install the kapp and secretgen controllers onto
our Kubernetes cluster
```

```
## Creating namespace tanzu-cluster-essentials
namespace/tanzu-cluster-essentials created
… much more output
Succeeded
ubuntu@ip-172-31-33-59:~/tanzu-cluster-essentials$
```

As the *kapp* CLI waits for the install to complete successfully, you'll see quite a lot of output to the console describing the progress of the install. If you get the Succeeded message at the end, you know that the installation did in fact succeed. Now that you have Cluster Essentials installed and the installation packages relocated, we're almost ready to install Tanzu Application Platform.

Setting up a developer namespace

The Tanzu Application Platform official documentation puts this step after the install, but I find that the install goes more smoothly if we do it first. This step involves creating a secret in a namespace where developers will deploy a workload as well as some Kubernetes RBAC artifacts so that the namespace's default service account can interact with the CRs that make up Tanzu Application Platform.

First, let's create the namespace and add a secret to it that will allow us to read from and write to our container registry. This example works for gcr.io, as we need to pass the password as a file. To create a secret for Azure, Docker Hub, or Harbor, you would use the --username and --password flags. My namespace-naming preference is workload1, workload2, and so on. This makes it clear to anyone perusing the cluster where the workloads are running:

```
ubuntu@ip-172-31-33-59 ~/t/packt> kubectl create ns workload1
namespace/workload1 created
ubuntu@ip-172-31-33-59 ~/t/packt> tanzu secret registry add
registry-credentials --server gcr.io --username _json_key
--password-file ./gcr-json-key.json -n workload1
| Adding registry secret 'registry-credentials'...
 Added registry secret 'registry-credentials' into namespace
'workload1'
ubuntu@ip-172-31-33-59 ~/t/packt>
```

Next, we need to give the default service account in our namespace some roles and secrets that will enable Tanzu Application Platform workloads to run in this namespace with the default service account:

```
ubuntu@ip-172-31-33-59:~/tap/packt$ cat <<EOF | kubectl -n
workload1 apply -f -
apiVersion: v1
kind: Secret
metadata:
```

```
  name: tap-registry
  annotations:
    secretgen.carvel.dev/image-pull-secret: ""
type: kubernetes.io/dockerconfigjson
data:
  .dockerconfigjson: e30K
---
apiVersion: v1
kind: ServiceAccount
metadata:
  name: default
secrets:
  - name: registry-credentials
imagePullSecrets:
  - name: registry-credentials
  - name: tap-registry
---
apiVersion: rbac.authorization.k8s.io/v1
kind: RoleBinding
metadata:
  name: default-permit-deliverable
roleRef:
  apiGroup: rbac.authorization.k8s.io
  kind: ClusterRole
  name: deliverable
subjects:
  - kind: ServiceAccount
    name: default
---
apiVersion: rbac.authorization.k8s.io/v1
kind: RoleBinding
metadata:
  name: default-permit-workload
roleRef:
  apiGroup: rbac.authorization.k8s.io
  kind: ClusterRole
```

```
    name: workload
subjects:
  - kind: ServiceAccount
    name: default
EOF
secret/tap-registry created
serviceaccount/default configured
rolebinding.rbac.authorization.k8s.io/default-permit-
deliverable created
rolebinding.rbac.authorization.k8s.io/default-permit-workload
created
ubuntu@ip-172-31-33-59:~/tap/packt$
```

Next, we'll proceed to install some tools that run on our Kubernetes cluster that will manage the installation for us.

Installing a package repository

Now we have all the client- and server-side tools ready to go, we can finally kick off our install. First, we need to set up a package repository. This is like a Docker or Apt repository. It's simply a service on the network that stores and serves software packages. In this case, we're installing a *Kapp* package repository on our Kubernetes cluster from which the Kapp controller can install packages. Previously, we relocated packages from the Tanzu registry to our own container registry. One of those packages is called `tap`. That `tap` package functions as our repository. It knows about all the other packages in the container registry. This will make more sense shortly. This repository and all its packages will live in their own namespace on our Kubernetes cluster. We'll assume that you still have your environment variables set from when we relocated the images. If they're not set, you'll need to go back and set them again:

```
ubuntu@ip-172-31-33-59:~/tap/packt$ kubectl create ns
tap-install
namespace/tap-install created
ubuntu@ip-172-31-33-59:~/tap/packt$ tanzu package repository
add tanzu-tap-repository --url ${INSTALL_REGISTRY_
HOSTNAME}/${INSTALL_REGISTRY_PROJECT_NAME}/tap-packages:${TAP_
VERSION} -n tap-install
  Adding package repository 'tanzu-tap-repository'


  Validating provided settings for the package repository


  Creating package repository resource
```

```
Waiting for 'PackageRepository' reconciliation for
'tanzu-tap-repository'

 'PackageRepository' resource install status: Reconciling

 'PackageRepository' resource install status: ReconcileSuc-
ceeded

Added package repository 'tanzu-tap-repository' in namespace
'tap-install'
ubuntu@ip-172-31-33-59:~/tap/packt$ tanzu package repository
get tanzu-tap-repository --namespace tap-install
NAME:         tanzu-tap-repository
VERSION:      789358
REPOSITORY:   gcr.io/fe-rhardt/tap-packages
TAG:          1.2.1
STATUS:       Reconcile succeeded
REASON:

ubuntu@ip-172-31-33-59:~/tap/packt$ tanzu package available
list --namespace tap-install


   NAME                                          DISPLAY-
NAME                                             SHORT-DE-
SCRIPTION
                                                 LATEST-VERSION
   accelerator.apps.tanzu.vmware.
com                     Application Accelera-
tor for VMware Tanzu
Used to create new projects and configura-
tions.                                           1.2.2
   api-portal.tanzu.vmware.com                   API
portal                                           A
unified user interface to enable search, discovery and try-out
of API endpoints at ease.


# … many more …
```

Now, we'll install these packages to stand up a running instance of Tanzu Application Platform.

Pulling down and formatting tap-values.yaml

Next, we need a config file to install Tanzu Application Platform. You can pull down this file as a starting point: https://github.com/PacktPublishing/DevSecOps-in-Practice-with-VMware-Tanzu/blob/main/chapter-09/tap-values.yaml.

Once you have that file locally, follow the inline instructions to plug in registry credentials and domain information. Once that file is filled out, we can install the tap package from our package repository, referencing that file:

```
ubuntu@ip-172-31-33-59:~/tap/packt$ tanzu package install
tap -p tap.tanzu.vmware.com -v $TAP_VERSION --values-file
tap-values.yaml -n tap-install

 Installing package 'tap.tanzu.vmware.com'

 Getting package metadata for 'tap.tanzu.vmware.com'

 Creating service account 'tap-tap-install-sa'

 Creating cluster admin role 'tap-tap-install-cluster-role'

 Creating cluster role binding 'tap-tap-install-cluster-
rolebinding'

 Creating secret 'tap-tap-install-values'

 Creating package resource

 Waiting for 'PackageInstall' reconciliation for 'tap'
\ 'PackageInstall' resource install status: Reconciling
```

This installation can take several minutes. Furthermore, it will sometimes tell you that it failed when in truth, a component was taking a little too long. One thing you can do to get a little more feedback on the install is to monitor the **PackageInstall** (**pkgi**) objects in Kubernetes:

```
ubuntu@ip-172-31-33-59 ~> kubectl get pkgi -n tap-install -w
NAME                             PACKAGE
NAME                                              PACKAGE
VERSION    DESCRIPTION           AGE
```

```
api-portal              api-portal.tanzu.vmware.
com                         1.0.24              Reconcile
succeeded    72s
appliveview-connector   connector.appliveview.tanzu.vmware.
com                 1.2.1       Reconcile succeeded    72s
appsso                  sso.apps.tanzu.vmware.
com                         1.0.0               Reconcile
succeeded    38s
buildservice            buildservice.tanzu.vmware.
com                     1.6.1               Reconcil-
ing          72s
cartographer            cartographer.tanzu.vmware.
com                     0.4.3               Reconcile
succeeded    38s
```

Congratulations! You have a running instance of Tanzu Application Platform. Now let's configure the remaining components necessary for you to interact with it.

Creating DNS records

Once you've installed Tanzu Application Platform, you'll want to be able to reach the tools and deployed applications from a web browser. This is where you go to your domain's DNS service and point it at your Tanzu Application Platform installation. First, let's find out the endpoint of our load balancer service and point a wildcard DNS entry at it:

```
ubuntu@ip-172-31-33-59 ~> kubectl get svc -n tanzu-system-
ingress envoy
NAME      TYPE            CLUSTER-IP      EXTER-
NAL-IP                      PORT(S)                     AGE
envoy     LoadBalancer    10.100.247.164    accd60a8feb-
5740b1ac24853ea17e546-817853481.us-west-2.elb.amazonaws.
com    80:30409/TCP,443:30100/TCP    10m
ubuntu@ip-172-31-33-59 ~>
```

In this case, the envoy service in tanzu-system-ingress has a hostname as it's EXTERNAL-IP. If I controlled mydomain.com, I would go into the DNS configuration settings for my domain and create a CNAME record, *.mydomain.com, and point it to the LoadBalancer host from the kubectl command. In you're using Azure or GCP, you might get an IPv4 address rather than a hostname. In that case, you'll need to create an A record rather than a CNAME.

Here's a real-world example. I did my Tanzu Application Platform install with the `*.packtinstall.k10s.io` domain. After configuring my DNS records properly, I can use the dig command to verify that any subdomain of `packtinstall.k10s.io` will return a CNAME pointing to the AWS load balancer associated with the `tanzu-system-ingress/envoy` Kubernetes LoadBalancer service:

```
ubuntu@ip-172-31-33-59 ~> dig testing123.packtinstall.k10s.io

; <<>> DiG 9.16.1-Ubuntu <<>> testing123.packtinstall.k10s.io
;; global options: +cmd
;; Got answer:
;; ->>HEADER<<- opcode: QUERY, status: NOERROR, id: 17729
;; flags: qr rd ra; QUERY: 1, ANSWER: 4, AUTHORITY: 0,
ADDITIONAL: 1

;; OPT PSEUDOSECTION:
; EDNS: version: 0, flags:; udp: 65494
;; QUESTION SECTION:
;testing123.packtinstall.k10s.io. IN A

;; ANSWER SECTION:
testing123.packtinstall.k10s.io. 300
IN    CNAME    accd60a8feb5740b1ac24853ea17e546-817853481.
us-west-2.elb.amazonaws.com.
accd60a8feb5740b1ac24853ea17e546-817853481.us-west-2.elb.
amazonaws.com.    59 IN A    52.10.176.193
accd60a8feb5740b1ac24853ea17e546-817853481.us-west-2.elb.
amazonaws.com.    59 IN A    35.155.17.135
accd60a8feb5740b1ac24853ea17e546-817853481.us-west-2.elb.
amazonaws.com.    59 IN A    54.213.143.29
```

Now, if everything worked perfectly the first time, you should be able to access the Tanzu Application Platform GUI with a browser pointed to `tap-gui.<your-domain>.com`. Success!

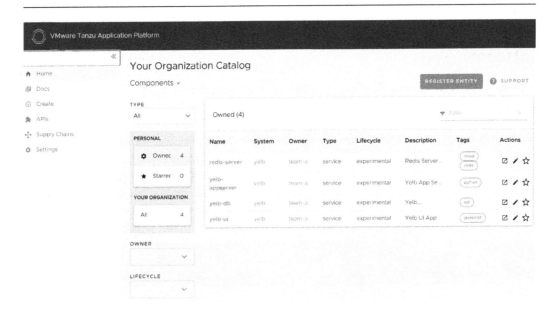

Figure 8.5 – Tanzu Application Platform GUI

I'd encourage you to explore the *TAP GUI*. *Chapters 2* and *5* of this book cover the Application Accelerator and API Portal components in detail. Next, let's install some local developer tools to enhance our experience interacting with Tanzu Application Platform.

Installing the VS Code developer tools

To really see Tanzu Application Platform in action, it's helpful to set up a true-to-life developer toolchain. At this point, you have a Kubernetes cluster running Tanzu Application Platform and you have `kubectl` and the Tanzu CLI with the necessary plugins. With just a few more additions, we'll have that toolchain.

First, if you don't already have it, download and install VS Code from here: `https://code.visualstudio.com/download`. Then, download and install Tilt from here: `https://docs.tilt.dev/install.html`. Finally, download the developer tools for VS Code from Tanzu Network here: `https://network.tanzu.vmware.com/products/tanzu-application-platform`. You install the developer tools from the VS Code Command Palette. You can open the Command Palette from the **View** menu. Then, start typing `Extensions: Install from VSIX`. Once the auto-complete comes up, select that option, and navigate to the VSIX file you just downloaded to install it.

Congratulations! You've completed the installation and configuration of a very complex piece of software. Our day-1 tasks are complete. Now, let's move on to day 2 and put our powerful new toolkit to use!

Day 2 – deploying applications to Tanzu Application Platform

In *Chapter 2, Developing Cloud-Native Applications*, we went in depth into Application Accelerator. To deploy an application to Tanzu Application Platform, we'll revisit that product. Navigate to the *TAP GUI* that you just finished installing. It should be at `tap-gui.<your-domain>`. Then, in the menu down the left-hand side, click on **Create**. Finally, select the accelerator called **Tanzu Java Web App**. When you select it, you'll be prompted for a name, which you can leave as is, and a container registry prefix. If you're using `gcr.io`, it might be something such as `gcr.io/<your-project-id>/ tanzu-java-web-app`. The value you provide will be used in the Tiltfile to tell the Tilt tool where to push your app's source code so it can be picked up by a Tanzu Application Platform supply chain. This will make more sense shortly. Here's a reminder of what the accelerator screen should look like:

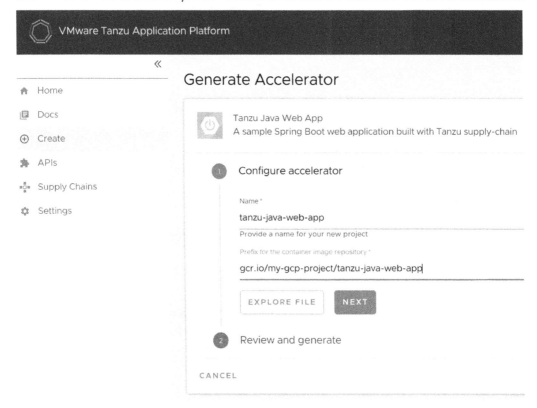

Figure 8.6 – Filling out the Accelerator form

Once you click **NEXT**, you'll shortly be presented with a button to download the ZIP file. Download the ZIP, unzip it, and open VS Code to the application directory you just unzipped. Then, we'll need to update a couple of settings in VS Code. Go into the VS Code Settings window and navigate to **Extensions | Tanzu**.

Set **Local Path** to the path of the unzipped application; or, if your VS Code file window is set to the root directory of the application, you can just leave this field blank. Set **Namespace** to `workload1` or whichever namespace you set up for developer use in the previous steps. Finally, double-check that **Source Image** is set to your preferred container registry. If you're using `gcr.io`, it might be `gcr.io/<your-project-id>/tanzu-java-web-app-remote-src`, depending on what you supplied to the Application Accelerator UI in the previous steps. Here's what my settings look like:

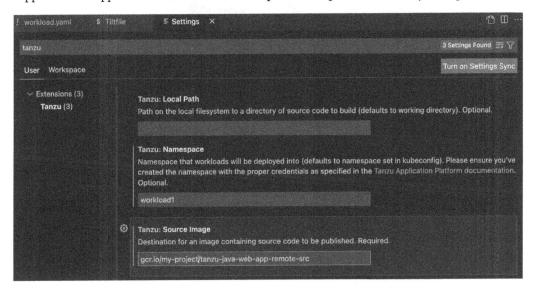

Figure 8.7 – VS Code developer settings

Now, you can close the **Settings** tab. In the **file** pane, look for a file called `Tiltfile`. When you right-click on the Tiltfile, you should see the **Tanzu: Live Update Start** option. The following screenshot gives you a good idea of what to look for:

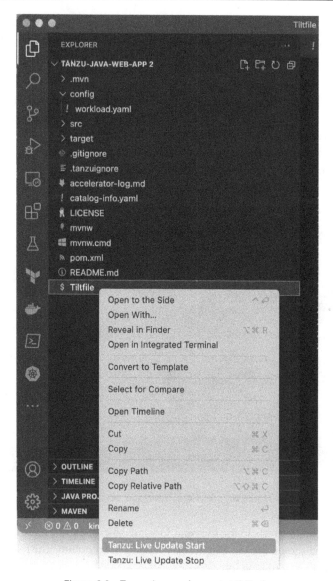

Figure 8.8 – Tanzu Java web app in VS Code

Let's try to start the live update by clicking on the **Tanzu: Live Update Start** context menu item. It's likely that you'll be met with a message like this:

```
*   Executing task: tilt: up - /Users/roberthardt/Downloads/
tanzu-java-web-app 2/Tiltfile

Tilt started on http://localhost:10350/
```

```
v0.26.3, built 2022-03-24

Initial Build
Loading Tiltfile at: /Users/roberthardt/Downloads/tanzu-java-
web-app 2/Tiltfile
ERROR: Stop! user@tap-install.us-west-2.eksctl.io might be
production.
If you're sure you want to deploy there, add:
        allow_k8s_contexts('user@tap-install.us-west-2.eksctl.
io')
to your Tiltfile. Otherwise, switch k8s contexts and restart
Tilt.
```

No worries. The Tilt tool is being extra cautious that you don't accidentally start a live update session in a production environment. Follow the instructions and add the `allow_k8s_contexts('<your-k8s-context>')` command to the bottom of the Tiltfile, then try again.

The first deployment may take a while, but eventually, you can open a terminal and look at all the pods in the workload1 namespace (`kubectl get pods -n workload1`), and see something like this:

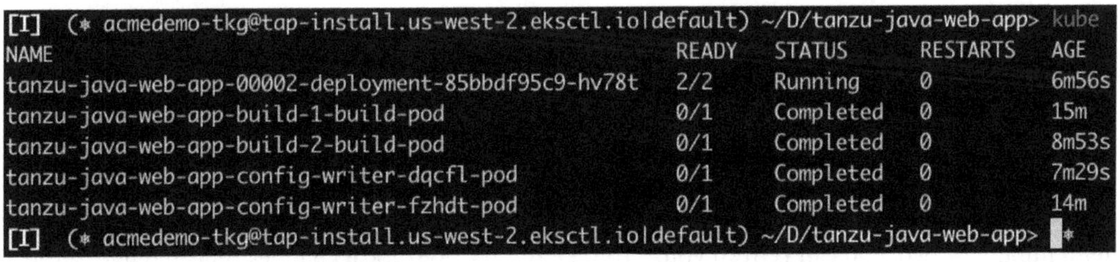

```
[I]  (* acmedemo-tkg@tap-install.us-west-2.eksctl.io|default) ~/D/tanzu-java-web-app> kube
NAME                                                   READY  STATUS     RESTARTS  AGE
tanzu-java-web-app-00002-deployment-85bbdf95c9-hv78t   2/2    Running    0         6m56s
tanzu-java-web-app-build-1-build-pod                   0/1    Completed  0         15m
tanzu-java-web-app-build-2-build-pod                   0/1    Completed  0         8m53s
tanzu-java-web-app-config-writer-dqcfl-pod             0/1    Completed  0         7m29s
tanzu-java-web-app-config-writer-fzhdt-pod             0/1    Completed  0         14m
[I]  (* acmedemo-tkg@tap-install.us-west-2.eksctl.io|default) ~/D/tanzu-java-web-app> *
```

Figure 8.9 – Pod listing of the running app

The pods containing the word `build` are those that take your source code and turn it into a container image. The pods containing `config-writer` are those that take the source for a deployable Kubernetes artifact and write it to Git or a container registry where it can be picked up and deployed. Finally, the pod containing `deployment` is your application's running pod.

Next, let's view our app in a browser. Execute this command to get the URL of the running Knative Service that belongs to your app:

```
~/D/tanzu-java-web-app> kubectl get kservice -n workload1
NAME                    URL                              LATESTCRE-
ATED                    LATESTREADY                      READY    REASON
tanzu-java-web-app      http://tanzu-java-web-app.workload1.pack-
tinstall.k10s.io    tanzu-java-web-app-00002    tanzu-java-web-
app-00002    True
```

Now, plug the URL into your browser. This is what you should see:

Greetings from Spring Boot + Tanzu!

Figure 8.10 – Running the Tanzu application

Congratulations! You have a Spring Boot application up and running on Tanzu Application Platform. Furthermore, there was no need to install a local Java runtime or any tools beyond the Tanzu toolchain. The plugin transported your raw source code onto the platform where it was built and deployed. Next, let's put the *Live Update* function to the test.

In VS Code, navigate to `/src/main/java/com/example/springboot/HelloController.java`, then change the greeting. Perhaps add `and Packt!!!` to the end.

Notice that when you save your change, the VS Code plugin automatically detects the change and deploys it. Here's a timeline of what happens when you make a change to a file in VS Code:

1. The Tilt plugin notices that a file has changed.
2. The plugin bundles up your source code and stores it in your container registry.
3. The plugin updates the workload object on your Kubernetes cluster such that the `.spec. source.image` field points to the updated source code image in the container registry.
4. The cartographer controller on Kubernetes notices the change in the workload and triggers a new iteration of the supply chain.
5. The supply chain pulls the source code, builds it, pushes the container image of the built artifact, and runs that artifact, applying multiple conventions in the process.

After a minute or two, you can refresh your browser and see your local changes reflected live in the app running in Kubernetes.

Deploying workloads directly

We've now seen the developer workflow where local source code is packaged up and stored in a container repository and workloads are created from that source. In a real-world scenario, though, local source code doesn't go straight from the IDE onto a platform. Rather, that code goes into source control, which then drives some sort of continuous integration onto the platform. Let's briefly explore that use case now.

In the `/config` directory of the app that we downloaded and unzipped, you should see a file called `workload.yaml`. This instructs the platform to pull the source code from this URL: `https://github.com/sample-accelerators/tanzu-java-web-app`. Tanzu Application Platform will watch that Git repository and whenever it detects a change, it will pull the latest source code and update the running workload. Let's try this out now.

First, to avoid namespace collisions, let's delete the existing workload. Let's stop Tilt by right-clicking on the Tiltfile and selecting **Tanzu Live Update Stop**. Then, let's delete our workload with this command:

```
~/D/tanzu-java-web-app> kubectl delete workload -n
workload1  --all
```

Next, there's a small bug in some versions of Tanzu Application Platform that we need to work around. Although the Git repo we're referencing is public, the platform still wants a secret to exist for authN. Let's create an empty secret as a workaround:

```
~/D/tanzu-java-web-app> kubectl create secret generic -n
workload1 git-ssh
```

Now, let's deploy the workload directly, taking special care to specify the correct namespace:

```
~/D/tanzu-java-web-app> kubectl apply -f ./config/workload.yaml
-n workload1
```

Then, we can watch the pods progress as the workload deploys:

```
~/D/tanzu-java-web-app> kubectl get pods -n workload1
NAME                                                         READY
STATUS        RESTARTS    AGE
tanzu-java-web-app-00001-deployment-557f89455b-
pg2td   2/2     Running       0             61s
tanzu-java-web-app-build-1-build-
pod                     0/1      Completed  0             2m21s
tanzu-java-web-app-config-writer-7kl4k-
pod                     0/1      Completed  0             74s
```

Finally, you can verify that your workload is visible in the browser at `tanzu-java-web-app.workload1.yourdomain.com`.

Now that we're practically experts at using the platform, let's have a look at what we might want to do next as platform operators.

Day 3 – common operational activities on Tanzu Application Platform

Thus far, we've installed a fully functional application platform, but in some ways, it still may feel like a toy. For instance, no enterprise is going to allow workloads to be deployed without TLS. Also, the platform is a bit of a black box that takes our source code and turns it into a running application. How can we know that what's running on the platform is the exact code we checked in? Going in depth into these topics is a bit beyond the scope of this book, but what I will do is describe some additional operational tasks and the problems they solve, and point you to some useful in-depth resources for more information.

Securing running workloads with TAP GUI with TLS

For an application platform in the enterprise, securing all web endpoints with TLS is non-negotiable. There are a few options that I'll list here. Some involve procuring the certificate outside of Kubernetes and supplying it directly, and others involve using an open source project called *cert-manager* (`https://www.jetstack.io/open-source/cert-manager/`), which gets installed automatically with Tanzu Application Platform. Here are some ways you might procure and deploy a TLS certificate into Tanzu Application Platform:

- Directly supplied with a third-party certificate (*DigiCert*, *Verisign*, etc.)
- Directly supplied with a self-managed CA (OpenSSL)
- Directly supplied with a corporate CA-signed certificate (e.g., *Nokia NetGuard*)
- Directly supplied with *Let's Encrypt*
- Via *cert-manager* with a self-managed CA
- Via *cert-manager* with *Let's Encrypt* (my preference)

As you can see, there are many options, which could easily double the length of this chapter. If you don't have a strong preference, I'd recommend *Let's Encrypt* + *cert-manager*. If you're managing your domain's DNS via a major cloud provider such as AWS Route 53, the process can be simple and painless.

Once you've decided how you'll obtain a certificate, you'll need to decide what **Subject Alternate Names (SANs)** you'll need to request. These are the URLs that your certificate will validate via TLS. If you use the conventions that ship with Tanzu Application Platform, I'd recommend these SANs. Let's say that the domain that we're using during the setup process is `mydomain.com`; then, the SANs you'd configure would be as follows:

- `*.mydomain.com`
- `*.workload1.mydomain.com`
- `*.workload2.mydomain.com`
- `*.workload3.mydomain.com`

This will give you TLS that will work for workloads deployed to the `workload1`, `workload2`, and `workload3` namespaces. There are ways to work around the limitation so you can deploy to any arbitrary namespace, but I'll refer you to the documentation for how to do that as well.

Now that we've decided on how we'll obtain a certificate and identified our SANs, I'll send you to this excellent blog post, which goes into detail on how to install Tanzu Application Platform with

TLS support: `https://tanzu.vmware.com/content/blog/tanzu-application-platform-install-with-tls-and-azure-ad`.

Enabling testing and scanning

Another non-negotiable for running enterprise software on a platform is that it needs to deploy with a guarantee that it has passed all its unit tests and that it has been scanned for known vulnerabilities. Tanzu Application Platform can provide both these guarantees with a bit of extra configuration.

First, regarding testing, there's no way that any platform could account for every testing framework across every programming language, so rather than moving forward with a half-baked attempt, the architects of Tanzu Application Platform decided to put automated testing in the hands of the app's developer via Tekton Pipelines.

Tekton (`https://cloud.google.com/tekton/`) provides Kubernetes-native continuous integration and delivery. Tanzu Application Platform, then, will look for a Tekton pipeline that knows how to run an application's tests and run it. The documentation also gives some sample pipelines for languages such as Java and Maven.

Scanning for vulnerabilities is a bit simpler. Tanzu Application Platform will scan an app's source code and generated container images using two well-known scanning tools: *Grype* or *Snyk*.

You can learn how to enable testing and scanning for your application supply chains here: `https://docs.vmware.com/en/VMware-Tanzu-Application-Platform/1.2/tap/GUID-getting-started-add-test-and-security.html`.

Once you've tackled these topics, you should have a thorough knowledge of Tanzu Application Platform and have a good intuition for what *good* looks like in the enterprise. What comes next? Let's brainstorm some next steps in the next section.

Next steps

Now that you're a veritable Tanzu Application Platform expert, where can you go next? Here are some thoughts:

- Deploy Tanzu Application Platform across multiple Kubernetes clusters with dedicated clusters for viewing, building, and running workloads

- Enable managed services (databases, message queues, etc.) that can automatically bind their credentials to workloads

- Enable custom GitOps workflows that require an approved pull request before deploying to production

- Build your own custom supply chains using the Cartographer tools (`https://cartographer.sh/`)

Summary

Tanzu Application Platform is likely the most complex piece of software we cover in this book and could benefit from a dedicated book of its own. Mastering this technology will make you indispensable to any enterprise looking to run software in the cloud at scale. With Tanzu Application Platform done, we're going to wrap up the second part of the book where we covered *running* applications on Tanzu, and transition to the next section, where we talk about *managing* them with tools such as Tanzu Mission Control, VMware Aria operations for Applications, and Tanzu Service Mesh. This wraps up our treatment of running applications on Tanzu. I encourage you to continue on to the next section where we cover managing those applications. We'll start with **Tanzu Mission Control**, a single point of control for all of your enterprise Kubernetes assets across multiple clouds.

Part 3 –
Managing Modern Applications
on the Tanzu Platform

This part will cover some important Tanzu tools that help to manage, secure, and proactively observe the container applications and the underlying Kubernetes platform for important day-2 activities of multi-cloud-based deployments.

This part of the book comprises the following chapters:

- *Chapter 9, Managing and Controlling Kubernetes Clusters with Tanzu Mission Control*
- *Chapter 10, Realizing Full-Stack Visibility with VMware Aria Operations for Applications*
- *Chapter 11, Enabling Secure Inter-Service Communication with Tanzu Service Mesh*
- *Chapter 12, Bringing It All Together*
- *Chapter 13, Appendix*

Managing and Controlling Kubernetes Clusters with Tanzu Mission Control

In the previous section of the book, we covered the tools in the Tanzu portfolio that help us run cloud-native applications. We covered how Harbor can provide a secure home for your container images and how we can run those images using Tanzu Kubernetes Grid, which provides a uniform user experience across public and private cloud infrastructure. Finally, we took a deep dive into the area of developer productivity to automate and secure the software supply chain from an idea to a running application in production.

This section is about managing cloud-native apps and their corresponding Kubernetes infrastructure. To begin this section, in this chapter, we will learn about managing, securing, and governing a fleet of Kubernetes clusters of any flavor and on any infrastructure using **Tanzu Mission Control** (**TMC**). Followed by that, we will cover VMware Aria Operations for Applications, a tool to monitor every part of your running applications, including distributed tracing between microservices, application performance, Kubernetes objects, virtual infrastructure, and other services used by the applications. Finally, we will learn how to connect your apps running on Kubernetes and deploy them in different clusters and environments securely with out-of-the-box mutual TLS configuration using Tanzu Service Mesh.

> **Sidenote**
> Henceforth in this chapter, we will refer to a *Kubernetes cluster* just as a *cluster* for brevity.

With the background from the last section and our forward-looking statement for the upcoming chapters in this section, let's begin our journey of understanding TMC in depth. We will cover the following topics in this chapter:

- *Why TMC?* – Understand the challenges around managing large Kubernetes environments and the solutions offered by Tanzu Mission Control

- *Getting started with TMC* – Learn how to start using TMC to manage the cluster lifecycle

- *Protecting cluster data using TMC* – Learn how to back up clusters and namespaces, and how to restore them when required to protect running workloads from disasters

- *Applying and ensuring governance policies on clusters using TMC* – Learn how to apply different cluster user governance policies to cluster and Kubernetes namespace groups and run inspections to find anomalies in clusters as a preventative security measure

TMC is a very powerful SaaS offering under the Tanzu portfolio that provides a single pane of control for all your Kubernetes environments. Let's learn more about it.

Why TMC?

With its increasing popularity, Kubernetes is the new infrastructure layer. Just as 10 years back, almost every software used to run on the virtual infrastructure, in the next few years, almost every new application will probably be deployed on Kubernetes by default. In fact, most of the vendor-provided solutions are now available to run on Kubernetes. As per a survey done by the **Cloud Native Computing Foundation** (**CNCF**) in December 2021, over 5.6 million developers said they used Kubernetes to deploy their applications, which is a 67% increase in just 1 year! Because of the array of business and technical benefits provided by Kubernetes, it is here to stay for a long time. Based on this CNCF survey, over 96% of organizations have embraced Kubernetes with a different level of maturity to run their cloud-native applications! And 73% of them have workloads running in production on Kubernetes already! As per a blog post by the CNCF in February 2022, the community is seeing its highest-ever adoption of this technology. Learn more about the details published by the CNCF here: `https://www.cncf.io/announcements/2022/02/10/cncf-sees-record-kubernetes-and-container-adoption-in-2021-cloud-native-survey/`.

Challenges with Kubernetes

Even though Kubernetes is an extremely popular tool to run containerized applications, operating Kubernetes is difficult in a production-grade environment. As you may know about Kubernetes, it only talks via the `kubectl` command-line interface or using its REST APIs. Every configuration in Kubernetes is in YAML, which is often long and difficult to understand unless you have a good grasp of different Kubernetes constructs.

Challenges with very large clusters

A large cluster, in this context, is a cluster with more than 50 nodes in it. Adding to the basic complexity of using Kubernetes, running a production-grade platform has several other security and operational challenges. To reduce complexity and operational overhead, I have seen enterprises deploy very large clusters to maintain, operate, and secure only a few critical clusters. Very large Kubernetes deployments host applications from several different **lines of business (LOBs)** using the logical isolation provided by Kubernetes namespaces. At first, this approach may sound logical, but there are several drawbacks to this approach, as listed in the following points:

- A large cluster serving multiple distinct LOBs and applications is very difficult to maintain, as all the applications and their LOBs will have different preferences in terms of the maintenance window and tolerance to any downtime. If an application is deployed with two or more Pods in a cluster, then it will mostly not face downtime. However, ensuring this level of compliance is a different challenge. Additionally, large clusters need a large maintenance window to complete activities such as upgrades.

- When there are hundreds of apps running in a large cluster, they all get impacted together if the cluster faces any issue or downtime. The large clusters have large blast radii and large disaster impacts.

- As a large cluster is used by several different applications, the application teams do not get any freedom of choice to deploy and run their apps with a specific cluster setup and resource requirements such as using a specific operating system or using a GPU for compute needs. Although Kubernetes provides a way to deploy certain application Pods on certain nodes using constructs such as taints and tolerations, implementing, maintaining, and using this kind of setup is practically very difficult.

- Although Kubernetes isolates applications of different teams using namespaces, this is only a logical level of isolation. In reality, Pods belonging to different namespaces may run in the same node. This may create issues of potential security threats posed by a malicious actor running in the same cluster and possible resource starvation because of a "noisy neighbor" if the application has not requested and reserved required resources as a part of its deployment manifest.

Considering these challenges around large clusters, it is recommended to use multiple smaller clusters, especially for heterogeneous workloads. If all the apps running on a large cluster have the same requirements, then it should be okay; otherwise, the recommended approach is to create smaller clusters for different teams, LOBs, or applications.

Challenges with many clusters and solutions from TMC

People generally prefer large clusters to avoid the operational overhead that increases in proportion to the number of clusters to be maintained. Hence, on one side, we may need several small clusters that belong to different teams, environments, and purposes, and on the other side, there are several operational and security challenges involved in keeping them functional. That is where TMC comes into the picture, which addresses these challenges by managing a fleet of clusters from a single pane of glass. Let's understand briefly what those challenges are and how TMC helps to solve them.

Increased overhead of cluster lifecycle management

Basic lifecycle management operations of several clusters, such as creating, scaling, upgrading, and deleting, could be a nightmarish situation without complete automation in place. Any manual intervention in this process could lead to configuration drifts resulting from basic human errors. This could soon result in a group of clusters having different configurations, and the need for cluster upgrades would be frequent for all the clusters, as the upstream Kubernetes releases a new version every 3 to 4 months and patch versions even more frequently. Regular maintenance of clusters via upgrades and patches is highly recommended to stay secure and supported. Such frequent maintenance of a large number of clusters requires a sophisticated automation setup. On the other hand, building and maintaining full end-to-end automation for these lifecycle processes require huge in-house effort. In that case, it's a question of the organizations building this automation asking themselves whether they should invest in this much effort for below-value-line activities to support a container platform, or whether they would rather invest these resources to add new business functionalities to their applications, which would bring more revenue.

TMC addresses this challenge with the help of **Tanzu Kubernetes Grid** (TKG). In the previous chapter, we learned about TKG and its concept of a management control plane to manage the lifecycle of several workload clusters under that management cluster. TMC provides an out-of-the-box integration experience to link TKG management control plane clusters and allows TMC users to perform all TKG cluster lifecycle operations using the TMC portal. Such an easy, quick, and user-friendly approach to managing cluster lifecycles addresses the challenges involved in keeping the Kubernetes versions up to date for security reasons. Additionally, a complete out-of-the-box cluster lifecycle automation offering from TKG and its integration with TMC reduces the additional toil required to create and maintain in-house automation for the same reason.

Distinct configuration requirements for different clusters

Things get even more complex when different clusters have different configuration requirements. The security requirements for a group of non-production clusters would not be the same as the production clusters. Furthermore, the compliance requirements of the environments handling **Payment Card Industry** (**PCI**) data are even more stringent. Depending on the environment, the clusters may have different needs for user access, container network, workload isolation, and deployment policies. A security policy such as preventing the deployment of privileged containers that may allow root-level access to the host they are deployed on, or a policy ensuring high availability only allowing two more Pods to be deployed for an application are just some examples of these kinds of policies. It is difficult to find that one size that fits all clusters. Adding more flavors of clusters adds increasingly more complexity.

To address this challenge, TMC allows us to create groups of clusters and then create different sets of policies and treatments for different groups of clusters. That way, the access policy of a development group of clusters could be different and more lenient than that of a group of production clusters. Additionally, TMC also allows you to create a group of Kubernetes namespaces, called **Workspaces**, which can span across the boundaries of clusters and cluster groups. Then, we can create policies for these Workspaces that are applicable at the namespace level. One such policy in Kubernetes is network policy defining, which applications/services may connect to which ones across different namespaces.

Installation and configuration of tools

Every organization has a set of tools that they want to deploy on their clusters for cross-cutting concerns such as logging, monitoring, certificate management, **identity and access management (IAM)**, and more. When different teams are in charge of the management of their own small clusters, it can become challenging to provide a self-service approach to the cluster owners where they can pick and deploy required tools on their clusters as and when required quickly and consistently.

To address this point, TMC has a catalog of some popular open source packages that different cluster owners may install based on their requirements with a single click. This capability not only reduces the overhead for different teams but also allows you to instate a guardrail for other cluster owners so that only certain authorized packages can be installed on their clusters. As an additional advantage, all these packages can be installed using a common approach, and only for the published versions in the catalog.

Cluster configuration and workload data protection

Access to a quick and easy procedure to back up and restore cluster states is also an important concern for most of the clusters in an organization. For static workloads that do not use persistent storage volumes, backing up the Kubernetes configuration YAML files may work to restore these objects by applying those configurations again, but what if there was a delta in the running state versus the documented state? The restored data, in this case, would not be identical. Backup and restoration of the stateful workloads add more complexity, as we need to also back up the storage volumes used by those workloads along with their configuration YAML files. Additionally, performing these activities for many clusters to make sure this happens at the required frequency to lose minimal data asks for huge automation efforts. On the other side, if the ownership of backup and restoration is left to the teams responsible for each individual cluster, then either the ball will be dropped or there will be a standardization challenge in how this is done across different teams.

To address this, TMC uses **Velero** (`https://velero.io`), an open source project backed by VMware, to back up and restore Kubernetes clusters. TMC users can either schedule cluster backups or do them on demand. These backups are saved in **Simple Storage Service (S3)**-compatible object storage locations accessible using web URLs. Taking, scheduling, and restoring backups using the TMC console is very easy and provides a consistent method and storage location to ensure the required data protection for everything running on a cluster, including stateless and stateful workloads. By means of Velero, TMC also allows you to only back up selected important namespaces if desired.

> **What is S3?**
>
> S3 is an object or file storage service offered by AWS. S3 has become a standard for object storage that implements the S3 interfaces required. Presently, there are several cloud-hosted and on-premises S3-compatible storage options available on the market in addition to the one offered by AWS, including MinIO and Dell **Elastic Cloud Storage (ECS)**.

Conducting cluster inspections

When an organization has hundreds of clusters owned by multiple teams, it is a mammoth effort to inspect each cluster in terms of its security and operational policy compliance. Enterprises operating in highly secure domains such as healthcare and finance have formal obligations to audit their application platforms at a regular frequency to find compliance and address violations. It is almost impossible to audit and inspect several different clusters manually to see whether they follow required security compliance practices. It requires a great amount of automation to build a scanning engine that can check all the different rules on a checklist for a given cluster and then report violations of different severities. There is a list of recommendations from the official Kubernetes specification that a secure production-grade cluster should follow. Similarly, the **Center for Internet Security (CIS)** also has a benchmarking list of recommendations that should be followed in a cluster for security. Conducting different types of inspections for hundreds of clusters is a very difficult endeavor.

However, running inspections for clusters only takes a few clicks using TMC. To provide this feature, TMC uses **Sonobuoy** (`https://sonobuoy.io`), another open source project backed by VMware, providing Kubernetes cluster configuration validation. TMC allows you to run two different types of inspection – CIS benchmarking and Kubernetes specification conformance. Upon the completion of these inspections, the TMC user gets a detailed report of the compliance and violation of different recommendations. These results can help you take quick preventative measures to close vulnerable security loopholes before they are exploited.

Along with solving these challenges around managing a large group of clusters, TMC also emits critical events related to the clusters under its purview to keep their owners fully informed about their health and critical lifecycle stages. Additionally, TMC offers a comprehensive set of REST APIs to allow you to perform all these operations programmatically. Finally, as a major benefit, TMC can perform all these operations for any conformant Kubernetes flavor, including AWS **Elastic Kubernetes Service (EKS)**, **Azure Kubernetes Service (AKS)**, **Google Kubernetes Engine (GKE)**, OpenShift, Rancher, open source upstream distributions, and many other types. This excludes the full cluster lifecycle management that we covered in the first point under *Increased overhead of cluster lifecycle management*. Upgrading Kubernetes clusters is only supported for TKG clusters at the time of writing. This capability makes TMC a great tool to implement a multi-cloud Kubernetes strategy for an enterprise. The following screenshot shows the TMC console with multiple Kubernetes clusters deployed on different cloud environments:

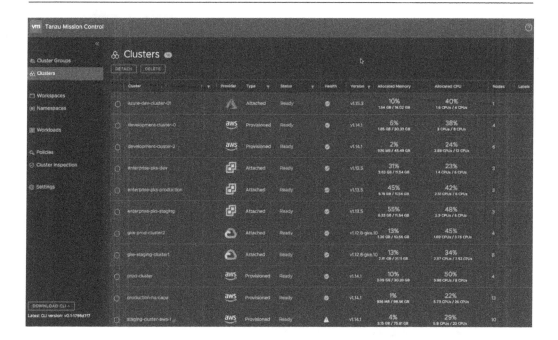

Figure 9.1 – TMC console with multiple Kubernetes clusters managed

With this, you should have gotten a convincible answer to the question, why TMC? We will get into the details of many of these capabilities later in this chapter, but for now, let's see how to get started with TMC.

Getting started with TMC

TMC is a **Software-as-a-Service** (**SaaS**) offering under VMware Cloud Services. Because of that, there is no installation and setup required to start using TMC, which is a big relief that SaaS products provide. In TMC, we can have two types of clusters, as described in the following points:

- **Clusters that are under a Kubernetes platform management control plane that is registered on TMC** – Presently, TMC only supports TKG clusters under this category. Once a TKG management cluster (a TKG platform control plane) running on vSphere, AWS, or Azure is registered in TMC, we can use the TMC interface to perform all lifecycle operations for all the clusters created under that management cluster.

- **Clusters that are attached to TMC** – These can be any conformant Kubernetes clusters that are created externally. We can attach these clusters to TMC for several management activities discussed previously in this chapter, except for full lifecycle operations, such as creating, deleting, and upgrading. TMC offers a common management control plane for all Kubernetes clusters irrespective of their flavors and vendors.

In this section, we will perform the following operations to get started with TMC:

- Accessing the TMC portal via the VMware Cloud Services console
- Registering an existing TKG management cluster running on AWS
- Creating a TKG workload cluster under the registered management cluster
- Attaching a GKE cluster for management
- Creating a cluster group
- Adding two associated clusters in the newly created cluster group
- Creating a Workspace, a group of cross-cluster Kubernetes namespaces
- Adding two Kubernetes namespaces from two different clusters to the created Workspace

However, before you can follow along, the following prerequisites must be fulfilled:

- Administrator-level access to TMC via a VMware Cloud Services account.
- An existing TKG management cluster (version 1.4.1 or later) with a production plan with at least three control plane nodes
- An existing Kubernetes cluster not managed by the previously mentioned TKG management cluster – this could be a GKE, AKS, EKS, OpenShift, Rancher, or even an open source Kubernetes cluster
- The cluster nodes should contain 4 vCPUs and 8 GB memory for the smooth execution of steps
- A user workstation with an internet browser (preferably Google Chrome) and the `kubectl` CLI
- Full internet connectivity from the Kubernetes clusters without any proxy servers, as the procedure to configure TMC for different operations differs with a proxy in between, which we are not considering in this chapter
- Access to either AWS S3 or any other S3-compatible object store service from the clusters linked with TMC.

Additionally, if your TKG management cluster is running on AWS and if you have configured the IAM permissions defined in the CloudFormation stack used by TKG manually, then you must add the following listed permissions to the `nodes.tkg.cloud.vmware.com` IAM policy or role:

```
{
  "Action": [
    "servicequotas:ListServiceQuotas",
    "ec2:DescribeKeyPairs",
    "ec2:DescribeInstanceTypeOfferings",
    "ec2:DescribeInstanceTypes",
```

```
    "ec2:DescribeAvailabiilityZones",
    "ec2:DescribeRegions",
    "ec2:DescribeSubnets",
    "ec2:DescribeRouteTables",
    "ec2:DescribeVpcs",
    "ec2:DescribeNatGateways",
    "ec2:DescribeAddresses",
    "elasticloadbalancing:DescribeLoadBalancers"
  ],
  "Resource": [
    "*"
  ],
  "Effect": "Allow"
}
```

However, these permissions are included automatically when you create or update the CloudFormation stack by running the `tanzu mc permissions aws set` command. Once these prerequisites are addressed, we should be good for the rest of the chapter. Let's start executing our plan to get started with TMC.

Accessing the TMC portal

Let's first open the TMC portal via the VMware Cloud Services portal with the following steps:

1. Visit the VMware Cloud Services portal at this URL: `https://console.cloud.vmware.com/`.

2. Click on the **LAUNCH SERVICE** link on the **VMware Tanzu Mission Control** tile as shown in the following screenshot:

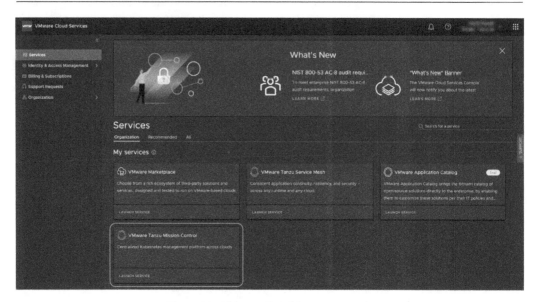

Figure 9.2 – VMware Cloud Services console

3.   That should open the TMC portal as shown in the following screen:

Figure 9.3 – Tanzu Mission Control console

Tip

If you are not entitled to TMC in your VMware Cloud Services account, then you will not see the TMC tile on the VMware Cloud Services console. You may need to request access for a TMC trial by getting in touch with your VMware contact point, or you can request access to TMC Starter via this URL: `https://tanzu.vmware.com/tmc-starter`.

Now, as we are on the TMC portal, let's register our first TKG management cluster in TMC.

Registering a TKG management cluster on TMC

Take the following steps on the TMC portal to register your existing TKG management cluster, running either on top of vSphere, AWS, or Azure cloud environments. The management cluster used in this chapter is deployed on top of AWS, which does not change any procedure in the following steps, except entering some configuration details for the workload cluster that we will create:

Registering a TKG management cluster on TMC

Take the following steps on the TMC portal to register your existing TKG management cluster, running either on top of vSphere, AWS, or Azure cloud environments. The management cluster used in this chapter is deployed on top of AWS, which does not change any procedure in the following steps, except entering some configuration details for the workload cluster that we will create:

1. Click on the **Administration** menu option from the left navigation bar and open the **Management clusters** tab as shown in the following screenshot:

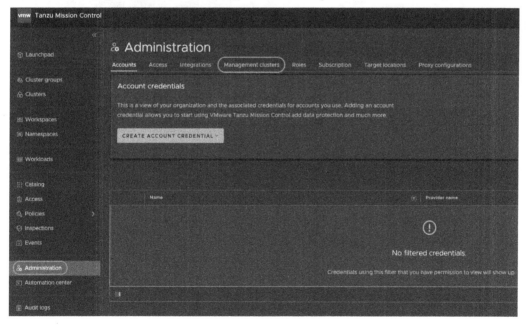

Figure 9.4 – Opening the Management clusters screen

2. On the **Management clusters** tab, click on the **REGISTER MANAGEMENT CLUSTER** dropdown, and select the **Tanzu Kubernetes Grid** option as highlighted in the following screenshot:

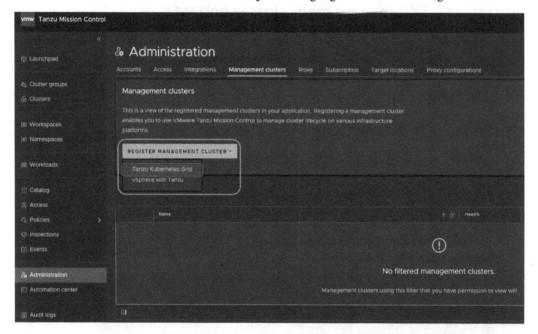

Figure 9.5 – Selecting Tanzu Kubernetes Grid to register as a management cluster

3. On the detail screen to register the management cluster, enter a unique name, select **default** from the drop-down list of groups, optionally add a small description, and finally, click on the **NEXT** button. We will discuss cluster groups in detail later in the chapter:

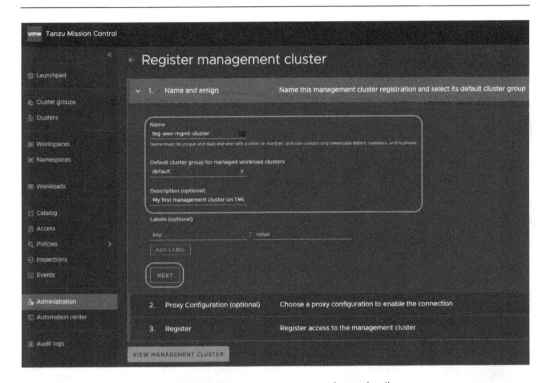

Figure 9.6 – Entering management cluster details

4. Click on the **NEXT** button in the **Proxy Configuration** section, as we will not need it:

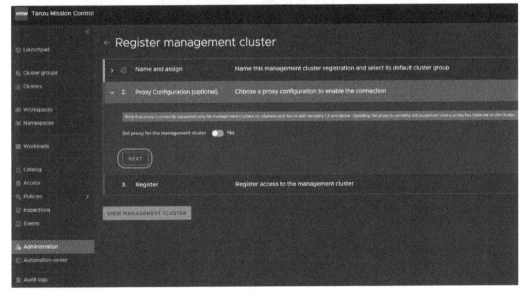

Figure 9.7 – Skipping proxy configuration

5. You will be given a URL of a YAML file that contains the Kubernetes resources that you need to create for your management cluster to link it with your TMC account. Copy the URL as shown in the following screenshot. You may expand the **View YAML** section to see the details of the Kubernetes resources that will be created on your management cluster and their configuration:

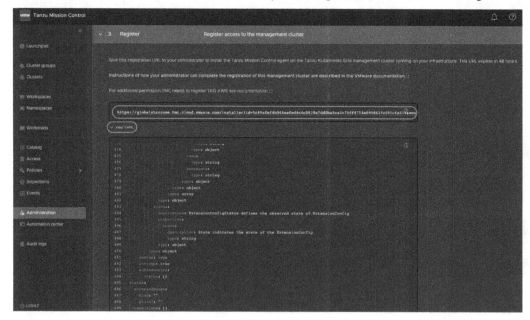

Figure 9.8 – Copying the registration URL to link to the management cluster

6. Click on the **VIEW MANAGEMENT CLUSTER** button to inspect the newly registered management cluster on TMC:

Figure 9.9 – Verifying the creation of the management cluster

7. You should see the page as displayed in the following screenshot. The status of the cluster is **Unknown**, as we are yet to apply the registration YAML configuration to our management cluster:

Figure 9.10 – Management cluster status unknown

8. Open your console window where you have access to the `kubectl` CLI on your workstation with the `kubectl` context pointing to the management cluster.

9. Run the following `kubectl apply` command using the URL copied in *step 4*:

```
$ kubectl apply -f <tmc-url>
```

This should list a bunch of different Kubernetes resources created for your management cluster. Once completed, we have created a two-way link between the TKG management cluster and the TMC account. We should be able to see the management cluster successfully verified on TMC.

10. After 5 to 10 minutes, either click on the **VERIFY CONNECTION** button as shown in *step 6*, or navigate to **Administration | Management clusters** to verify the successful registration of the management cluster. As you can see in the following screenshot, the cluster is **Healthy** and in a **Ready** state now. Click on the cluster name link as highlighted in the following screenshot:

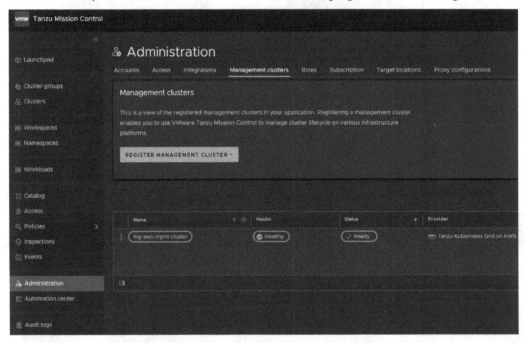

Figure 9.11 – Management cluster registration verification

11. Once you click on the name of the management cluster as shown in the previous step, you will see something similar to the following screenshot showing the details of the management cluster, including the health indicators of different components running on the cluster:

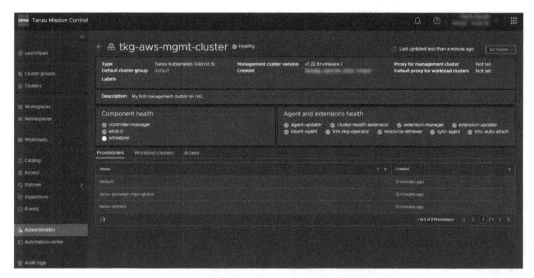

Figure 9.12 – Management cluster details

This concludes our steps to register a TKG management cluster in a TMC account using the TMC portal. In the next section, we will see how to use the TMC portal to create a new TKG workload cluster under the newly added management cluster.

Creating a new workload cluster under a management cluster

After setting up the first TKG management cluster on TMC, let's use this setup to create a new TKG workload cluster. As discussed in the previous chapter covering TKG in detail, a TKG workload cluster is used to run your containerized apps, whereas the management cluster is the control plane for several of these workload clusters.

All TKG workload clusters are created under a **provisioner**. A provisioner is a namespace within the management cluster that owns required workload clusters within the management cluster. This way, the provisioners in TKG provide a multi-tenancy construct to allow different teams to create and manage their own workload clusters without interfering with others. In our case, we will use the **default** namespace of the management cluster, which should be present by default.

Take the following steps to create a new TKG workload cluster for your TKG management cluster:

1. Go to the **Administration** menu of the left-hand navigation bar, open the **Management clusters** tab, and click on the **tkg-aws-mgmt-cluster** link to open its detail page:

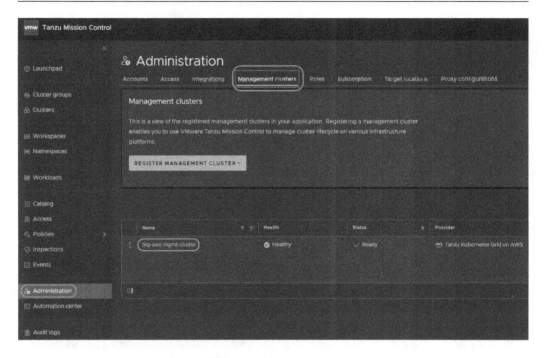

Figure 9.13 – Opening the management cluster page

2. Go to the **Clusters** menu on the left-hand navigation bar and click on the **CREATE CLUSTER** button as highlighted in the following screenshot:

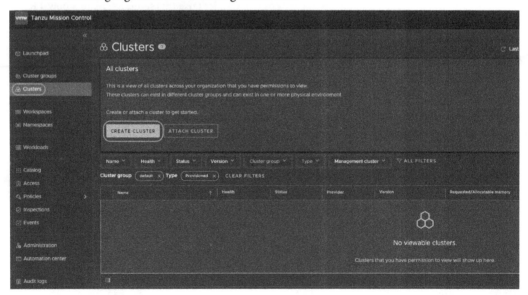

Figure 9.14 – Creating a workload cluster

3. Select **tkg-aws-mgmt-cluster** from the list that was registered on TMC earlier and click on the **CONTINUE TO CREATE CLUSTER** button:

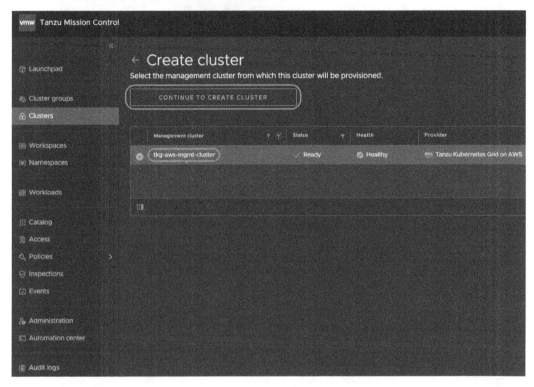

Figure 9.15 – Continue creating a workload cluster

4. Select **default** from the **Provisioner** dropdown and click on **NEXT**:

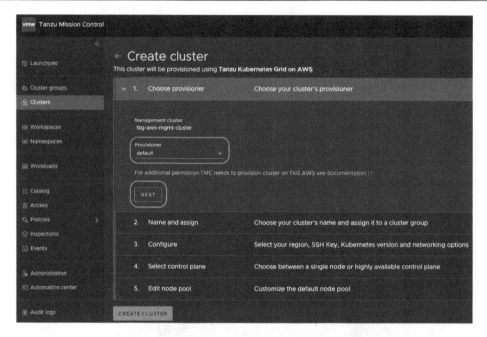

Figure 9.16 – Selecting the provisioner for the workload cluster

5. Enter the cluster name, optionally add a description, and click on the **NEXT** button:

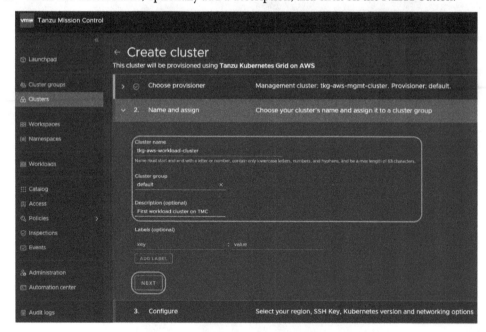

Figure 9.17 – Entering the workload cluster details

6. Enter the required infrastructure-specific details in this step and click on the **NEXT** button. As the TKG management cluster used in this chapter is running on AWS, the following screenshot shows AWS-specific details. For vSphere and Azure, some fields will be different. You can get more details about them here: `https://docs.vmware.com/en/VMware-Tanzu-Mission-Control/services/tanzumc-using/GUID-42150344-CD4C-43AE-8C39-C059A97EF47C.html`:

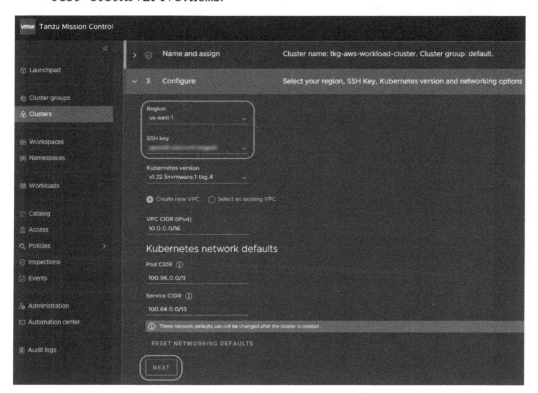

Figure 9.18 – Entering the workload cluster configuration

7. Choose the control plane plan for the workload cluster to be created. You can also optionally change the values of the other fields, but this is not required and we recommend following the procedure in this chapter. Finally, click on the **NEXT** button to select the worker node details:

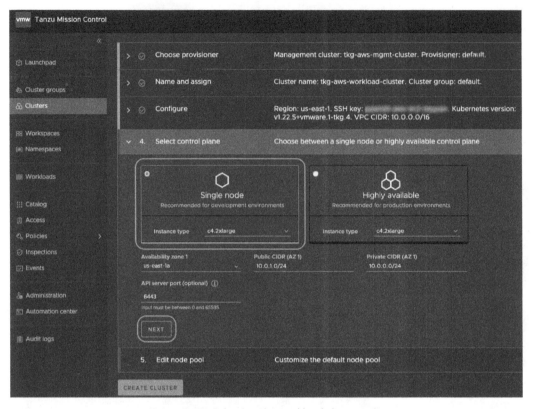

Figure 9.19 – Selecting the workload cluster type

8. Enter the count of the worker nodes for the workload cluster and click on the **CREATE CLUSTER** button:

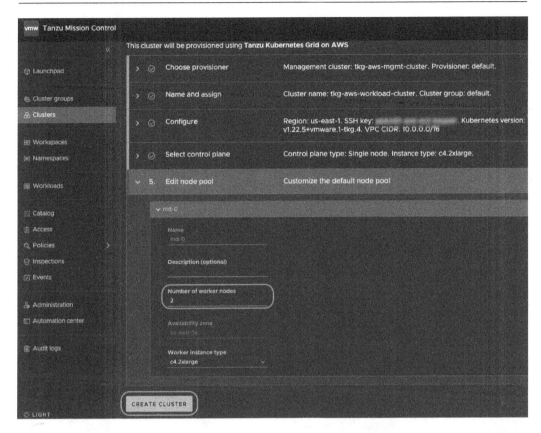

Figure 9.20 – Entering the worker node count

You will see a message saying that the workload cluster is being created with its specifications as shown in the following screen:

Figure 9.21 – Workload cluster being created

9. You will see a cluster detail screen like the following screenshot once the workload cluster is created successfully. It might take 5 to 10 minutes depending on the size and infrastructure:

Figure 9.22 – Workload cluster detail page

This concludes our third step, creating a TKG workload cluster using a TKG management cluster using the TMC portal. Now, we will attach an externally managed Kubernetes cluster to TMC so that we can perform various day-2 activities for that cluster using TMC.

Attaching an existing Kubernetes cluster with TMC

In this section, we will attach an existing GKE cluster to the TMC account. You may select any other type of Kubernetes cluster, including Rancher, AKS, EKS, OpenShift, upstream open source, and more. TMC will allow you to attach a Kubernetes cluster as far as it is a conformant Kubernetes cluster with admin-level `kubectl` access for the cluster. So, let's attach an external cluster:

1. Go to the **Clusters** menu from the left-hand menu bar and click on the **ATTACH CLUSTER** button as shown in the following screenshot:

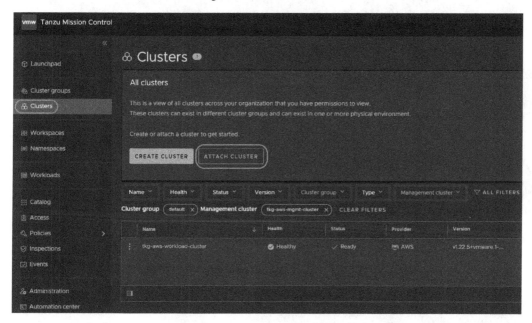

Figure 9.23 – Going to the Attach cluster page

2. Enter the cluster name, select the cluster group as **default**, optionally enter a description, and finally, click on the **NEXT** button as highlighted in the following screenshot:

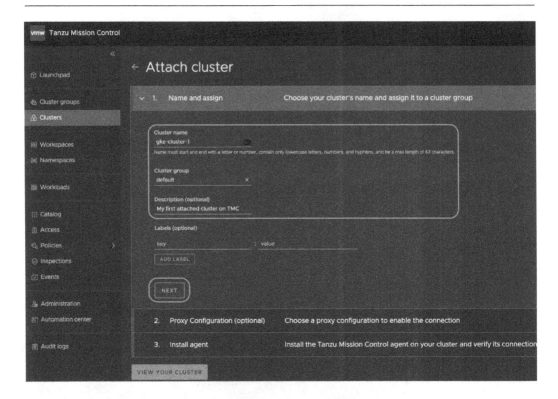

Figure 9.24 – Entering the attached cluster details

3. Skip the proxy configuration and click on the **NEXT** button:

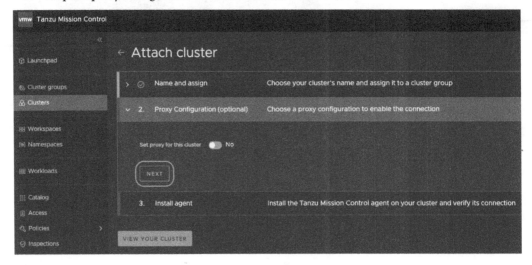

Figure 9.25 – Skipping the proxy configuration details

4. Just as we registered the management cluster using a Kubernetes YAML configuration file earlier in this chapter, it is now time to register the external cluster using a similar approach. You can view the details of the Kubernetes resources that will be created on the targeted cluster to establish two-way communication between the cluster and TMC:

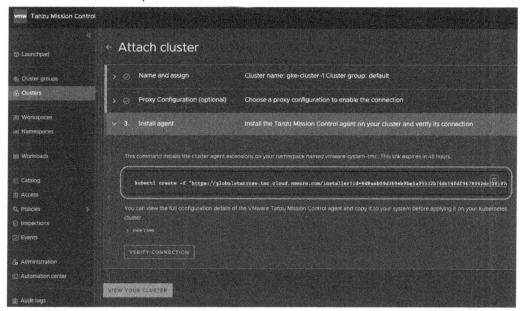

Figure 9.26 – Copying the agent configuration command to attach the cluster to TMC

5. Open the command window where you have access to the cluster being attached via `kubectl` and change the `kubectl` context to point to the cluster being attached.

6. Run the `kubectl` command copied in *step 4* on your command window to create the required agent deployment resources in your cluster to attach to TMC.

7. After the successful execution of the `kubectl create` command, click on the **VERIFY CONNECTION** button. You should see the recently attached cluster in the list as shown in the following screenshot. Then, click on the attached cluster name to verify its details:

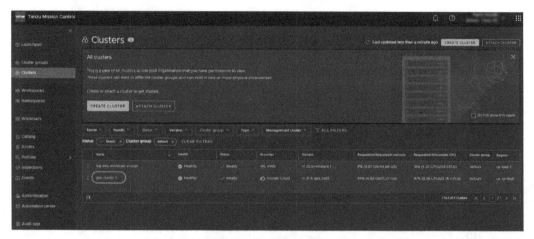

Figure 9.27 – Verifying an attached cluster in the list

8. Examine the details of the attached cluster. As you can see in the following screen, the type of the cluster is **Attached**:

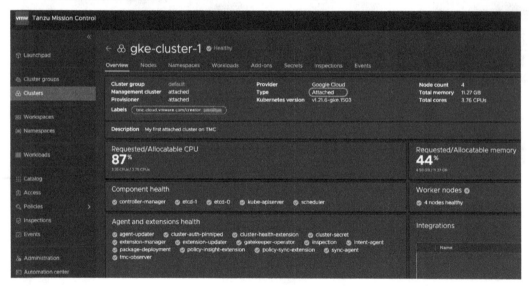

Figure 9.28 – Attached cluster details

This concludes our task of attaching an externally managed Kubernetes cluster to TMC. As the next steps, we will create a cluster group and add our TKG workload and GKE-attached clusters to that group.

Creating a cluster group on TMC

As discussed earlier in this chapter, TMC allows you to group different clusters of similar natures and manage and handle them using common configurations. Configuring groups of several clusters makes the operation of large Kubernetes foundations very easy and efficient. So, let's learn how to create cluster groups in TMC in this section:

1. Click on the **Cluster groups** menu item from the left-hand navigation bar and click on the **CREATE CLUSTER GROUP** button as shown in the following screenshot:

Figure 9.29 – Creating cluster groups

2. Enter the cluster group name, add an optional description, and click the **CREATE** button as shown in the following screenshot:

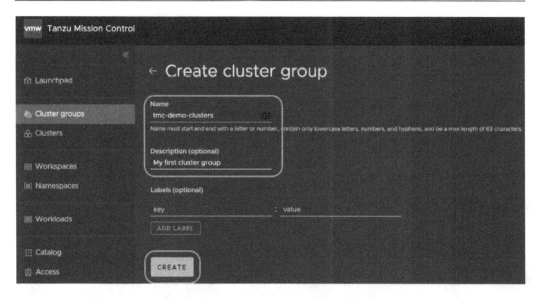

Figure 9.30 – Entering cluster group details

3.   You should see the newly created cluster group in the list as shown in the following screenshot:

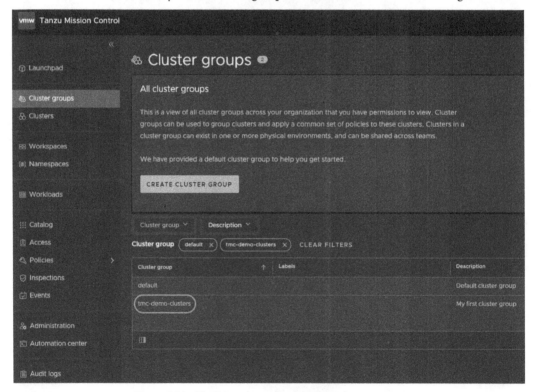

Figure 9.31 – Verifying the presence of the cluster group

Now, let's add our two clusters to this newly created group using the following steps.

4. Click on the **default** cluster group as shown in the following screenshot:

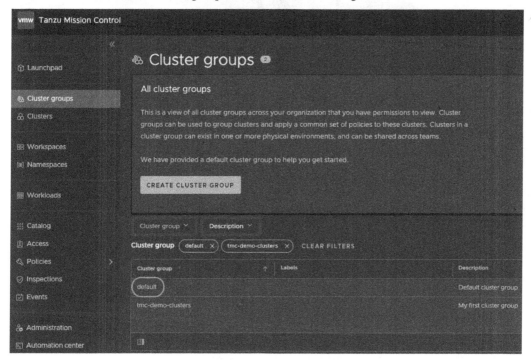

Figure 9.32 – Selecting the cluster group as default

5. Select the two clusters we have on TMC, for the TKG workload and GKE, and click on the **MOVE** button as shown in the following screenshot:

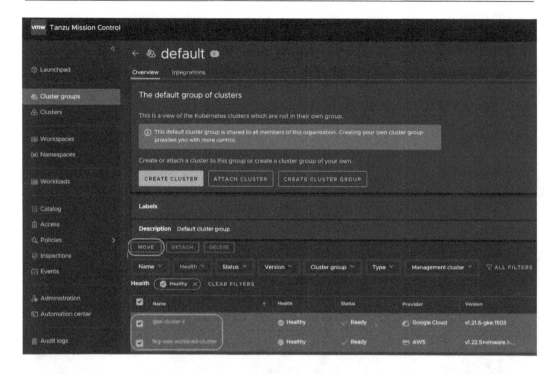

Figure 9.33 – Selecting clusters for grouping

6. Select the cluster group we created in *step 2* and click on the **MOVE** button:

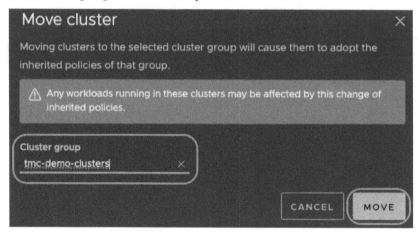

Figure 9.34 – Selecting the cluster group

7. Go to the **Cluster groups** page and click on the new cluster group we created to examine the presence of the two clusters we added in the previous step:

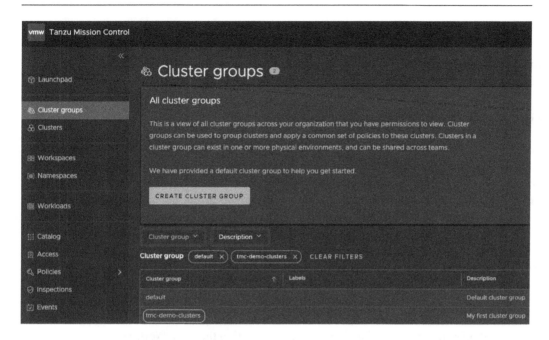

Figure 9.35 – Getting into the new cluster group

8. As you can see in the following screenshot, both clusters that we created in this section are in the cluster group:

Figure 9.36 – Listed clusters in the cluster group

Grouping clusters with similar requirements is a very powerful way to manage hundreds of clusters with heterogeneous requirements. Along with groups of clusters, TMC also allows you to group Kubernetes namespaces to perform certain namespace-level configurations applicable to more than one namespace in different clusters. Let's learn more about it.

Understanding Workspaces in TMC

As we discussed before in this chapter, TMC allows you to group clusters of similar natures. Grouping helps us create a common configuration and policies for them for operational efficiency. However, certain policies in Kubernetes can only be applied at the namespace level, such as network policies that define which Pods from one namespace can talk to which Pods in another namespace. Additionally, we can also configure cluster user access policies at the namespace level, and TMC also allows us to create image registry access policies at the namespace level. While applying cluster-level policies to a cluster group is an easy task, applying namespace-level policies for individual namespaces could be very impractical. Moreover, a way to apply namespace-level policies to a group of namespaces within a cluster would also not help much if we were trying to manage hundreds of Kubernetes clusters. For these reasons, TMC offers a construct called **Workspaces**, which allows us to create a group of namespaces across different clusters – then, we can create a namespace-level policy applicable to the entire Workspace and hence for all the namespaces within the same Workspace. A multi-cloud application that is deployed in different clusters might need similar configurations for either user access or network policy configuration. For a Workspace containing all the namespaces of this application and running different clusters, it would come in very handy to treat them as a unit.

Let's create two namespaces in the two clusters we have added in the previous section of this chapter and then create a Workspace to add those two namespaces into:

1. Open the **Workspaces** menu from the left-hand navigation bar and click on the **CREATE WORKSPACE** button as shown in the following screenshot:

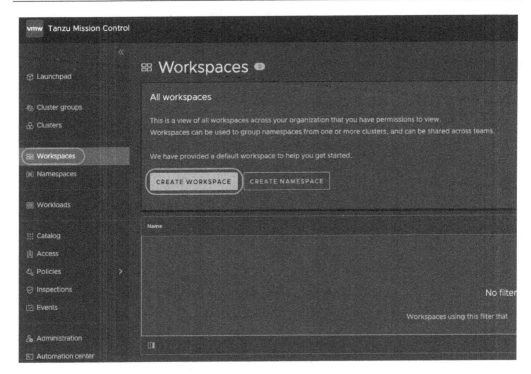

Figure 9.37 – Creating a Workspace

2. Enter a name, optionally add a description, and click on the **CREATE** button on the page as shown in the following screenshot:

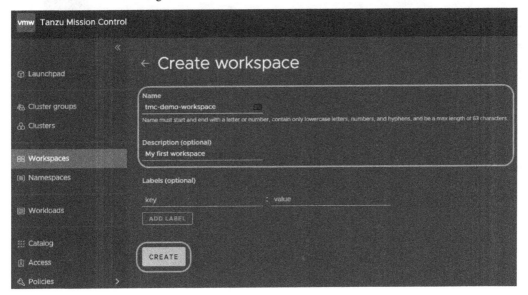

Figure 9.38 – Entering Workspace details

3. As you can see in the following screenshot, a new Workspace has been created. Click on the Workspace name to ensure there are no namespaces listed in there:

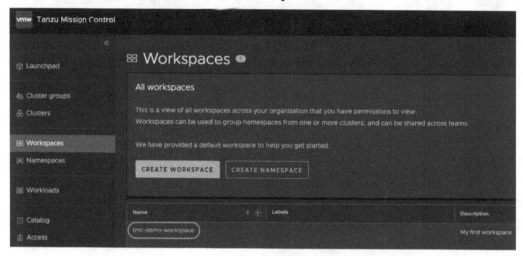

Figure 9.39 – Verifying that the Workspace has been created

4. Open the **tmc-demo-clusters** cluster group that we previously created on the TMC portal as shown in the following screenshot:

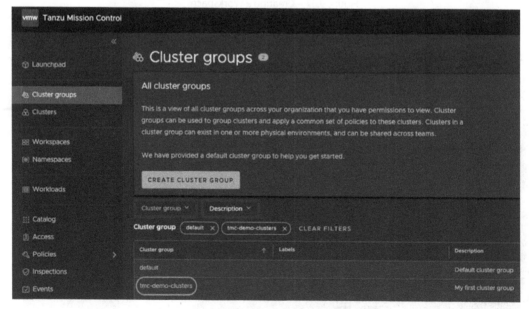

Figure 9.40 – Opening the cluster group

5. Click on the TKG workload cluster as shown in the following screenshot:

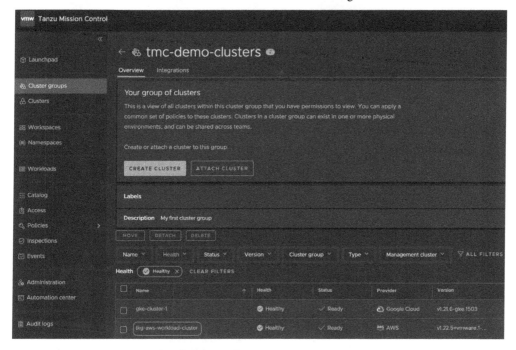

Figure 9.41 – Opening a cluster detail page

6. Click on the **Namespaces** tab of the cluster detail page, select the **default** namespace, and click on the **ATTACH 1 NAMESPACE** button as shown in the following screenshot:

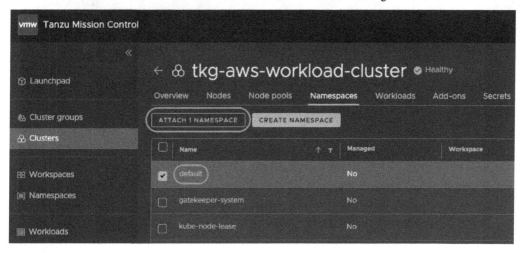

Figure 9.42 – Selecting a namespace to attach to a Workspace

7. Select or enter the Workspace name and click on the **ATTACH** button as shown in the following screenshot:

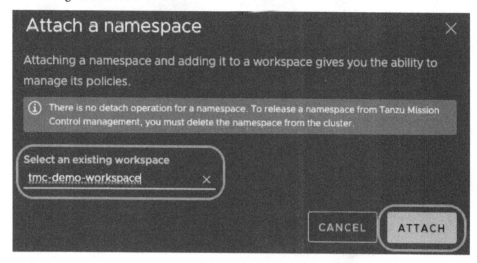

Figure 9.43 – Selecting the Workspace to attach

8. As you can see in the following screenshot, the **default** namespace is now under **tmc-demo-workspace**, which we created in the previous step:

Figure 9.44 – Verifying the namespace-Workspace association

9. Repeat *steps 5* to *8* from this list to add the **default** namespace to this new Workspace for the other cluster, **gke-cluster-1**, as per this chapter, which we previously attached to TMC and added to the **tmc-demo-clusters** group.

10. Open the **Workspaces** menu from the left-hand navigation bar, and click on the new Workspace, **tmc-demo-workspace**, that we created earlier:

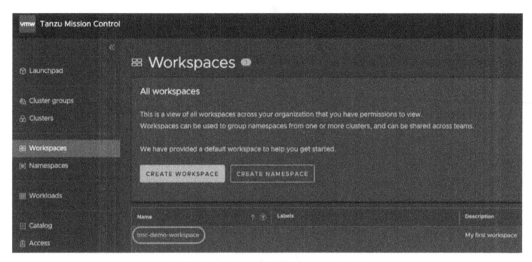

Figure 9.45 – Opening Workspace details

11. You should be able to see two default namespaces from two different clusters listed as a part of the new Workspace we created in this part as shown in the following screenshot:

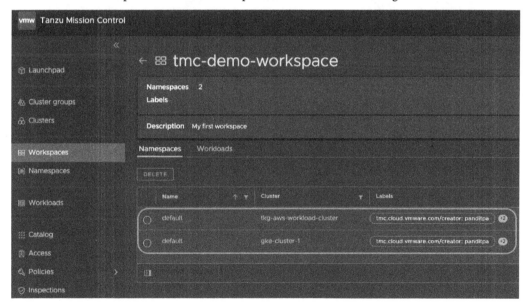

Figure 9.46 – Verifying namespaces in the Workspace

With this, we have covered how to create Workspaces and how to attach existing namespaces to different clusters in the Workspace. In the example that we covered earlier, we used the default namespaces with the same name as the clusters that were in the same cluster group in TMC. However, you can add any namespaces to a Workspace irrespective of the namespace names and the group status of their parent clusters.

With this, we conclude all the steps that we planned to cover under this part of the chapter – *Getting started with TMC*. We learned how to register a TKG management cluster in TMC and created a TKG workload cluster using the TMC interface. Then, we also attached an externally managed GKE cluster to TMC. Finally, we learned how to group clusters and Kubernetes namespaces to perform common operations on them as a unit. In the next part of the chapter, we will learn about making and restoring cluster data backups using TMC.

Protecting cluster data using TMC

Kubernetes is widely used to run business-critical applications in production environments. In these cases, a reliable disaster recovery mechanism should be present to make regular cluster data and configuration backups and restore them in the event of data loss for any reason. Although Kubernetes is mostly used to run stateless workloads where the persistent data is stored outside the clusters in the databases, running stateful software, such as caches, queues, and databases, is also being adopted slowly. In *Chapter 6, Managing Container Images with Harbor*, the Harbor registry deployment used the data stores that were deployed on Kubernetes itself. That makes backing up data even more important.

So, to cover this important topic, we will learn how to make cluster backups and restore them using TMC with the following high-level steps:

1. Configure an S3-compatible remote backup storage location.
2. Configure a cluster to use the remote storage location for backup data.
3. Deploy a custom application to the cluster.
4. Take a backup of the cluster.
5. Delete the cluster namespace with the custom deployment to simulate a disaster.
6. Restore the cluster backup.
7. Verify the presence of the deleted namespace and the custom deployment within it post-restoration.

Let's start the work to execute these steps one by one.

Configuring the backup target location

TMC allows you to create separate backup target locations for separate cluster groups. Configuring a backup location for a cluster group is a one-time administrative activity. Once a target location is associated with a cluster group, we can make either on-demand or scheduled backups for any cluster in the cluster group using the backup location configuration. Additionally, one configured backup location can be used by many cluster groups. In this chapter, we will use the AWS S3 option as the storage location.

The following steps describe how to configure a backup target location for a cluster group in TMC:

1. Create your AWS account credentials for TMC to use to store backup data in an S3 bucket:

    I.   Open the **Administration** menu from the left-hand navigation bar and click on the **CREATE ACCOUNT CREDENTIAL** button as shown in the following screenshot:

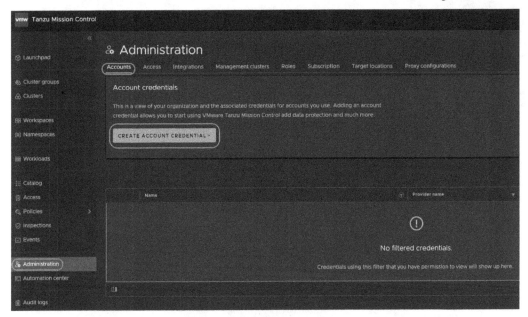

Figure 9.47 – Creating AWS account credentials

    II.   Select the **AWS S3** option from the **TMC provisioned storage** option:

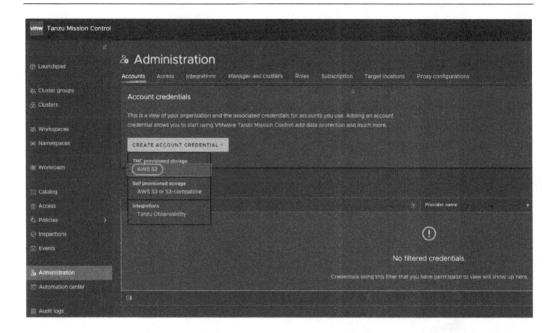

Figure 9.48 – Selecting provisioned storage

III.   Enter a name for the account credentials and click on the **GENERATE TEMPLATE** button:

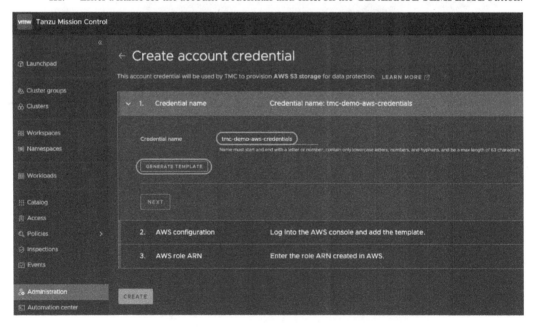

Figure 9.49 – Entering a credential name and generating a template

IV.   A credential template file should be downloaded. This is an AWS CloudFormation template that creates the required S3 buckets and grants required permissions to TMC to access that bucket for backup and restoration operations. Save that template file on your local workstation and click on the **NEXT** button:

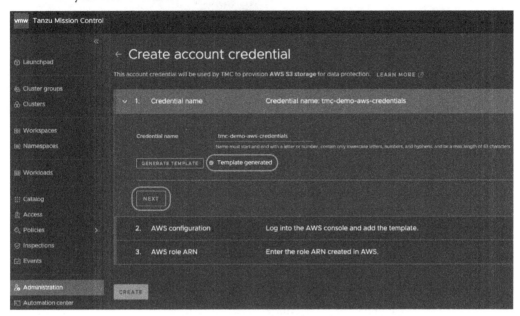

Figure 9.50 – Downloading the generated AWS CloudFormation template

V.   Follow the **QUICKSTART GUIDE** link to apply the downloaded template to your AWS account. Once the CloudFormation template is applied, you will get the **Amazon Resource Name (ARN)** in the output of the template execution, which will be required in the next step. You will find these instructions in the quick start guide as shown in the following screenshot:

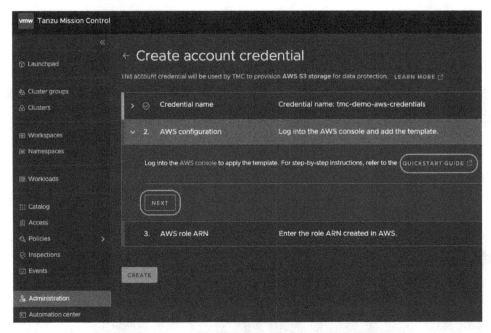

Figure 9.51 – Following the quick start guide to create the required AWS S3 objects

VI.   Apply the copied ARN from AWS after applying the CloudFormation template and click the **CREATE** button to finally create the account credentials that we can use for the backup and restoration operations for any cluster:

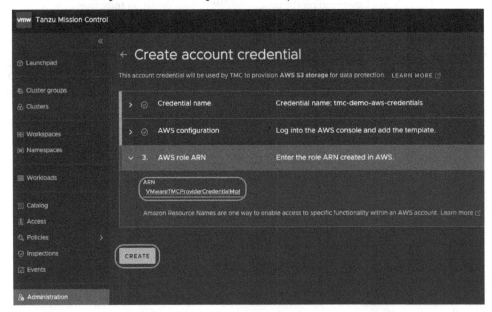

Figure 9.52 – Applying the copied ARN and creating the credentials

VII. Upon successful creation of the account credentials, you should see it listed under the **Accounts** tab of the **Administration** menu as shown in the following screenshot:

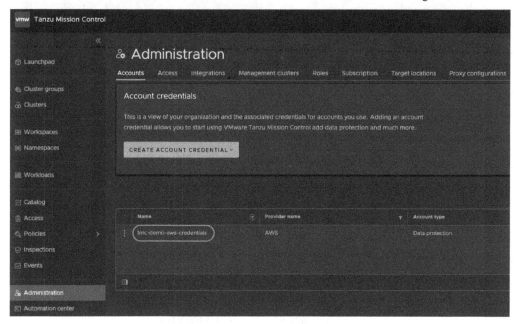

Figure 9.53 – Verifying account credential creation

2. Go to the **Target locations** tab under the **Administration** menu and click on **CREATE TARGET LOCATION** for the AWS S3 option as per the following screenshot:

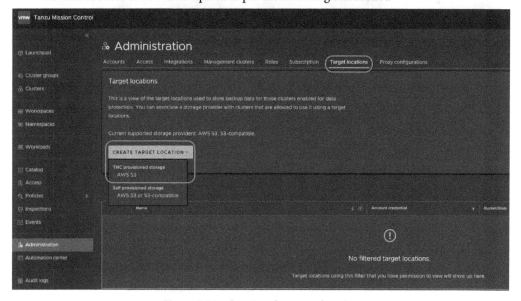

Figure 9.54 – Creating the target location

3. Select the account credentials that we created in *step 1* of this section and click on the **NEXT** button:

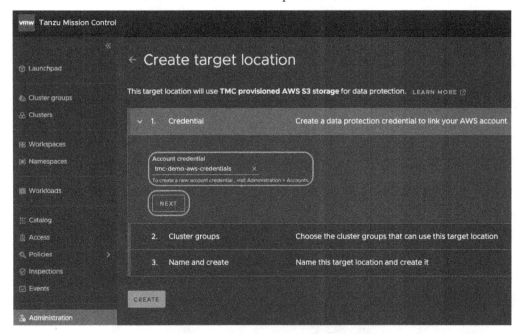

Figure 9.55 – Selecting account credentials

4. Select the cluster group that we created earlier in this chapter and click on the **NEXT** button:

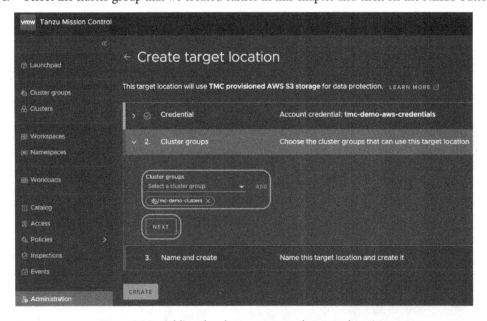

Figure 9.56 – Adding the cluster group to the target location

5. Enter the name of the target location and click on the **CREATE** button:

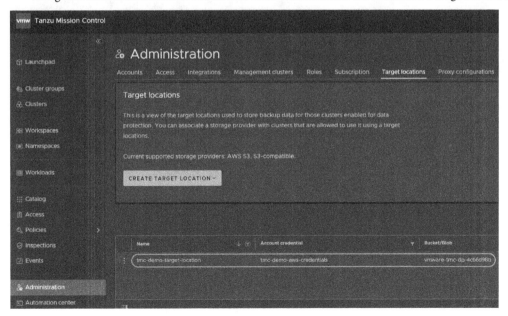

Figure 9.57 – Entering the target location name

6. You should see a new target location created under the **Target locations** tab of the **Administration** menu along with the account credentials and the relevant S3 bucket name as shown in the following screenshot:

Figure 9.58 – Verifying target location creation

Now, as we have set up a backup target location and assigned it to the cluster group, we can proceed to perform the backup and restoration operations for a cluster in that group.

Enabling data protection for a cluster

After completing the steps required to perform backup and restoration for a cluster group, let's enable it for one of the clusters in the group. In this example, we will use the externally managed GKE cluster that we had attached with TMC. Take the following steps to complete this task:

1. Click on the attached cluster, **gke-cluster-1**, in the **tmc-demo-clusters** group. If you have used names other than the ones used in this chapter, you will need to select the cluster appropriately:

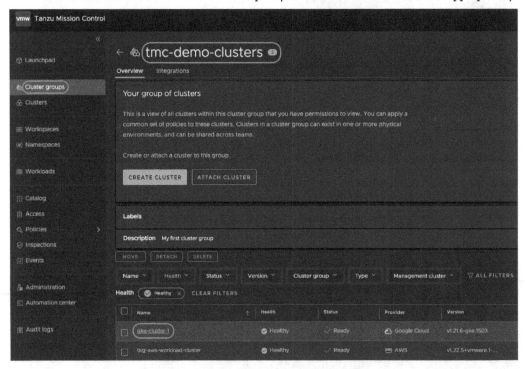

Figure 9.59 – Opening the attached cluster details page

2. Click on the **ENABLE DATA PROTECTION** link as highlighted in the following screenshot:

Figure 9.60 – Clicking on the ENABLE DATA PROTECTION link

3.   Click on the **ENABLE** button to confirm the operation:

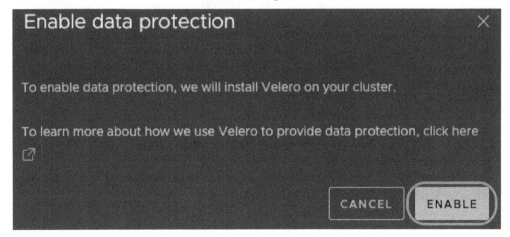

Figure 9.61 – Enabling data protection for the cluster

This operation deploys Velero, an open source Kubernetes cluster backup and restoration tool. With this, we have prepared the cluster to take backups with the required toolset deployed on it. Now, let's deploy a test workload in the cluster.

Deploying a custom application in the cluster

Execute the following steps to run an nginx deployment on the cluster we enabled for backup and restoration in the previous task:

1. While in the targeted cluster's `kubectl` context, run the following command, which creates a namespace named `nginx` and a Kubernetes deployment named `nginx-deployment` with three Pods running. You can check the deployment manifest file used in the following command to get more details:

    ```
    $ kubectl apply -f https://raw.githubusercontent.com/
    PacktPublishing/DevSecOps-in-Practice-with-VMware-Tanzu/
    main/chapter-10/nginx-deployment.yaml
    ```

2. Verify the deployment using the TMC portal under the **Workloads** tab of the cluster details page. As you can see in the following screen, the nginx deployment and ReplicaSet are running with a **Healthy** status. To minimize the clutter on the screen, you can also filter Tanzu and Kubernetes-specific workloads using the switches as highlighted:

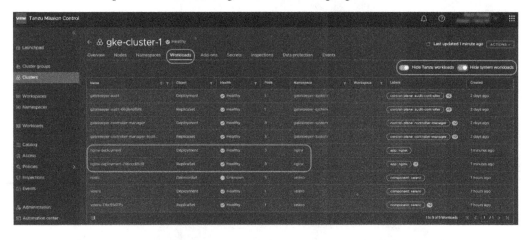

Figure 9.62 – Verifying workload deployment for a cluster

Now, we have a workload deployed in its own namespace on a target cluster that can be backed up. Let's make a backup of the cluster using the following steps.

Backing up a cluster

Take the following steps to take a backup of the cluster where we deployed the nginx workload in the previous task:

1. On the cluster's **Overview** tab, click on the **CREATE BACKUP** link as highlighted in the following screenshot:

Figure 9.63 – Initiating the cluster backup process

2. Select the option to make a backup of the entire cluster and click on the **NEXT** button. We can also make a backup only of selected namespaces or selected objects identified using a label value, which is a very flexible choice:

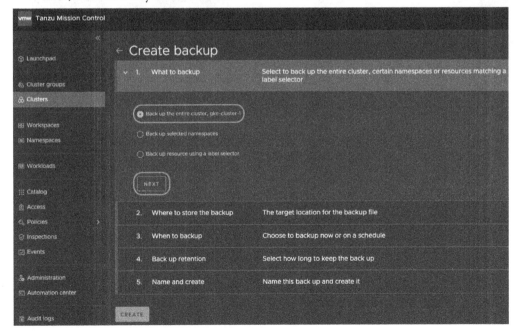

Figure 9.64 – Selecting the backup scope

3. Select the backup target location that we created previously in this chapter pointing to an AWS S3 bucket and click on the **NEXT** button:

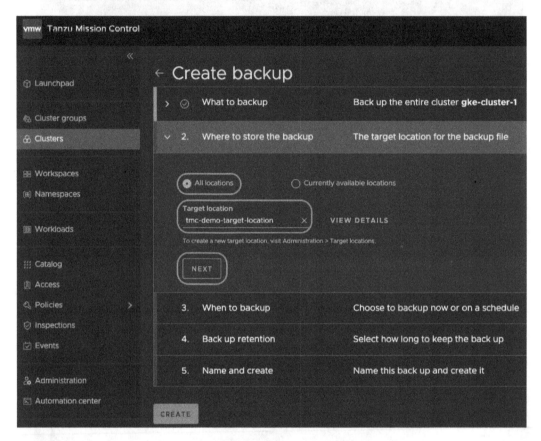

Figure 9.65 – Selecting the backup target location

4. Select the backup schedule. While we can create a regular backup schedule, here, we will select **NOW** to make an on-demand backup to learn about the concept. Click on the **NEXT** button to move on after that:

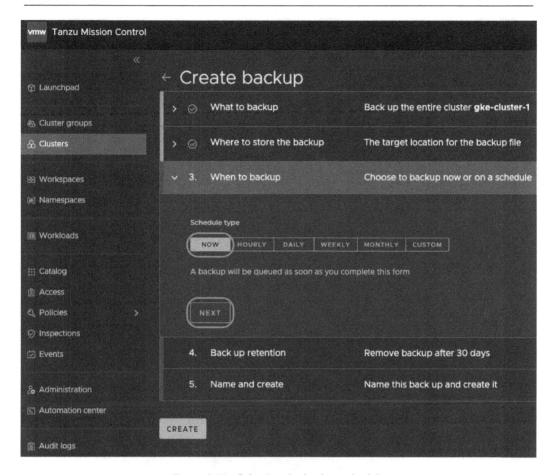

Figure 9.66 – Selecting the backup schedule

5.   Enter the backup **Retention (days)** value and click on the **NEXT** button:

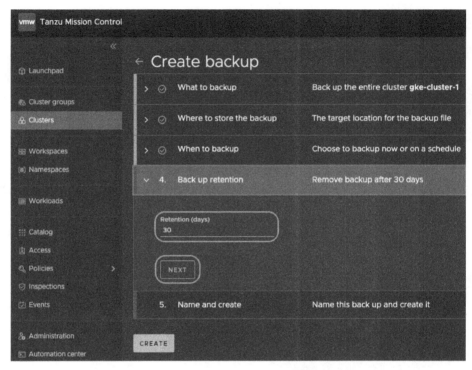

Figure 9.67 – Entering the backup retention days

6. Finally, enter the backup's name and click on the **CREATE** button as shown in the following screenshot:

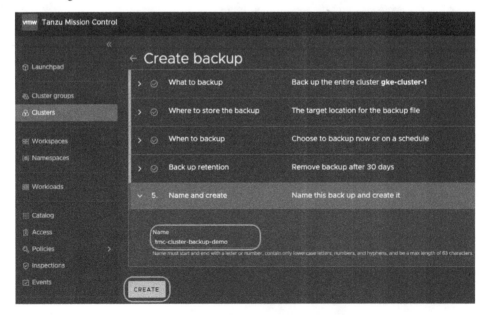

Figure 9.68 – Entering the backup name

7. This will trigger the backup process and within 2 to 5 minutes, you should be able to see the backup completed under the **Data protection** tab of the cluster as shown in the following screenshot. The backup process may take more time if the cluster has other workloads running:

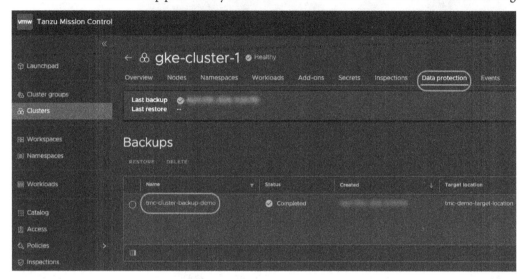

Figure 9.69 – Verifying backup completion

After successfully making a backup of the cluster, let's now restore it – but before we restore it, let's *accidentally* delete the **nginx** namespace that contains the nginx deployment we created earlier before making the backup.

Deleting a custom deployment running on the cluster

Take the following steps to perform this task:

1. Run the following command to delete the **nginx** namespace from the targeted cluster:

    ```
    $ kubectl delete namespace nginx
    ```

2. Verify the absence of the nginx deployment in the **Workloads** tab of the cluster. As you can see, **nginx-deployment** and its corresponding ReplicaSet that we verified previously are now missing:

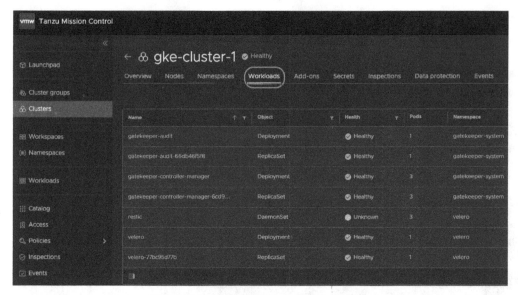

Figure 9.70 – Verifying the workload deletion

Now, as we have *accidentally* deleted the entire namespace and the workloads running in that namespace to simulate a disaster situation, let's use the backup we made for the cluster to restore the deleted namespace and get its objects back up and running.

Restoring the cluster backup

Take the following steps to restore the backup of the targeted cluster and bring back the deleted Kubernetes objects in that cluster:

1. Go to the **Data protection** tab of the cluster, select the backup that we took previously from the **Backups** list, and click on the **RESTORE** link as highlighted in the following screenshot:

Figure 9.71 – Restoring a backup

2. Since we had only one namespace deleted, we will just restore that one by selecting the highlighted option to restore specific namespaces. Select the **nginx** namespace from the list of what we need to restore. Here, TMC also allows us to restore the source namespace from the backup to a different target namespace if that is intended. This can be done using the little pencil icon beside the target namespace caption. After selecting the namespace, click on the **NEXT** button:

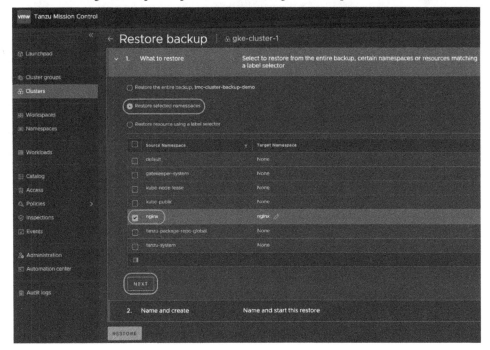

Figure 9.72 – Selecting the Restore backup specification

3.   Enter a name for the restored instance and click on the **RESTORE** button:

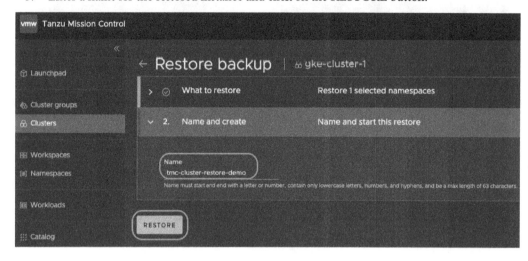

Figure 9.73 – Entering a restore instance name

4.   Upon successful restoration, you will see a restored entry in the **Data protection** tab of the cluster under the **Restores** section as shown in the following screenshot:

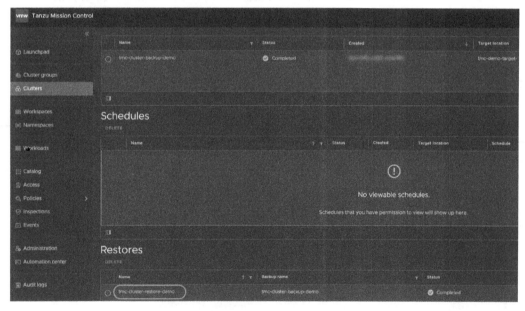

Figure 9.74 – Verifying that the restoration has completed

5.   Verify the presence of the **nginx** namespace and its objects under the **Workloads** tab of the cluster details page. As you can see, the **nginx** namespace has now been restored successfully:

Figure 9.75 – Verifying the presence of the restored objects

With this, we can conclude our very long but important section on backing up and restoring Kubernetes clusters. It is worth noting that several configurations made in this section were one-time activities. This includes creating AWS S3 account credentials, creating a target location of the backup, and associating a target location with a cluster group, among other things. Once this setup is done, backing up and restoring a cluster or a part of the cluster only takes a few clicks.

Let's now look into another very important capability of TMC – policy management.

Applying governance policies to clusters using TMC

In the previous section of the chapter, we learned how to get started with TMC by registering a TKG management cluster, creating a TKG workload cluster, attaching a GKE cluster, and finally, grouping them – but why do we bring all the clusters to TMC? In this section, we will check this out by performing various activities with these clusters using the TMC interface. We will cover the following activities:

- Configuring a security policy for a cluster group
- Configuring an image registry governance policy for a Workspace
- Configuring a deployment governance policy for a cluster group

- Checking policy violation status for clusters

- Inspecting a cluster for CIS benchmark compliance

This is a long list of activities to cover in this section. Let's knock them off one by one.

Configuring a security policy for a cluster group

When it comes to running containers, several things can be misconfigured from a security point of view, which keeps the door open for hackers to leverage these. Depending on the nature of the workloads running on the cluster, a Kubernetes administrator may need to secure several things. In this world of microservices, the Kubernetes platform team often needs to allow different teams to deploy their apps with their own configurations required by the apps. However, ensuring that all the teams follow the required security practices outlined by the platform team can be a very difficult task. That is why Kubernetes administrators need to guardrail their clusters so that they do not allow workloads to be deployed in unsecured ways. To address this need, Kubernetes offers a construct called **PodSecurityPolicy**, which defines what a Pod can and cannot do in the Kubernetes cluster with a PodSecurityPolicy in effect. TMC allows you to configure these security policies for a group of clusters so that there is no chance of configuration drifts between clusters.

Let's create one such security policy for the cluster group that we previously created. This policy will prevent any Pod from gaining privileged access to the Kubernetes node's operating system and resources. The following steps will help create and test this policy:

1. Ensure that a privileged access Pod can be deployed before applying the policy, taking the following substeps:

    I.   Deploy a privileged Pod in the TKG workload cluster under the `tmc-demo-clusters` group using the following command. You can check the Pod definition file used in the following command – notice that the security context is defined there to enable the Pod to get privileged access:

    ```
    $ kubectl apply -f https://raw.githubusercontent.com/
    PacktPublishing/DevSecOps-in-Practice-with-VMware-Tanzu/
    main/chapter-10/privileged-pod.yaml
    ```

    II. Verify whether the Pod has been created and is running successfully. Here, it is assumed that there is no security policy in place that would prevent a privileged Pod from being created:

    ```
    $ kubectl get pod
    NAME             READY    STATUS     RESTARTS    AGE
    privileged-pod   1/1      Running    0           48s
    ```

2. After verifying that we can run a privileged Pod in the cluster, let's now create a security policy in TMC for the cluster group using the following substeps:

    I.   Open the **Policies** > **Assignments** menu from the left-hand navigation bar and the **Security** tab. Then, click on the cluster group, **tmc-demo-clusters**, and click on the **CREATE SECURITY POLICY** button as highlighted in the following screenshot:

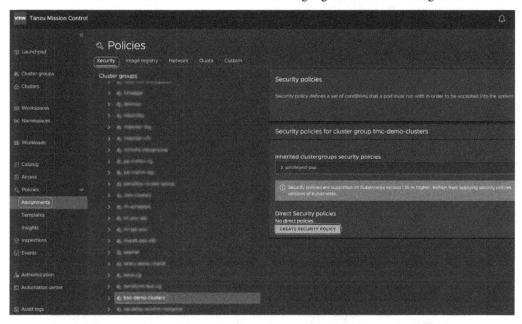

Figure 9.76 – Creating a security policy for a cluster group

    II. TMC allows you to fully customize your security requirements by defining a custom security policy from scratch. However, it also provides two out-of-the-box choices: **Baseline** and **Strict**. Let's select **Baseline** from the **Security template** dropdown. You will notice that it restricts the creation of privileged containers as highlighted in the following screenshot. Finally, click on the **CREATE POLICY** button to apply the restrictions to the relevant cluster group:

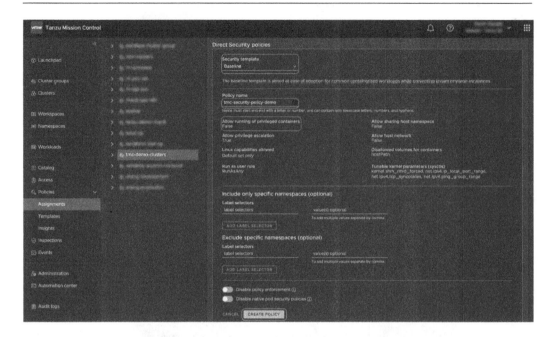

Figure 9.77 – Entering the security policy configuration

III.   You will see the policy listed for the cluster group as shown in the following screenshot:

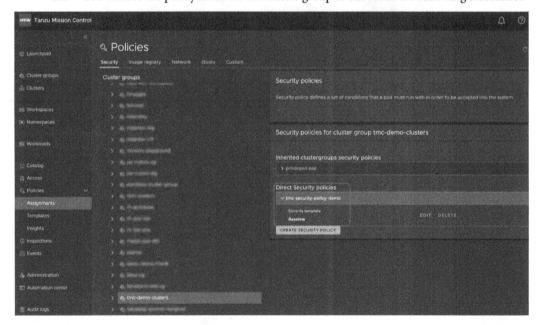

Figure 9.78 – Verifying security policy creation

3. Let's now verify whether the policy has been applied to the clusters in the group. Take the following substeps to do so:

    I.   First, delete the Pod we created in the first step in the TKG workload cluster by running the following command:

    ```
    $ kubectl delete pod privileged-pod
    ```

    II. Create the same Pod again by running the following command:

    ```
    $ kubectl apply -f https://raw.githubusercontent.com/
    PacktPublishing/DevSecOps-in-Practice-with-VMware-Tanzu/
    main/chapter-10/privileged-pod.yaml
    ```

    III. You will see an error message as follows, explaining that the Pod could not successfully be created because of the security policy in place:

    ```
    Error from server ([tmc.cgp.tmc-security-policy-
    demo] Privileged container is not allowed: centos,
    securityContext: {"privileged": true}): error
    when creating "https://raw.githubusercontent.com/
    PacktPublishing/DevSecOps-in-Practice-with-VMware-Tanzu/
    main/chapter-10/privileged-pod.yaml": admission webhook
    "validation.gatekeeper.sh" denied the request: [tmc.cgp.
    tmc-security-policy-demo] Privileged container is not
    allowed: centos, securityContext: {"privileged": true}
    ```

That concludes our learning on how to configure a Kubernetes cluster security policy applicable to deploying and running workloads using TMC. We tested one of the clusters in the cluster group to which we applied the policy, but you can also do the same exercise for another cluster in the group and should see similar test results. Like the one explained here for running privileged containers, we can create several different types of security policies using TMC for a cluster group. Let's now learn how to apply a Workspace-level policy that defines how the workloads running in the specific cluster namespaces can pull container images from a container registry.

Configuring an image registry policy for a Workspace

When it comes to pulling container images from a container registry, several proven practices are recommended. The restrictions applicable to pulling images for regulatory environments could be more stringent. The following are the parameters that TMC allows you to configure for a container registry policy depending on different compliance requirements. A registry policy in TMC may either have all or some parameters applicable as required:

- **Pulling images only using a digest (SHA) and not with a tag** – This is an important rule to set up for a production environment, as a tag could technically have different content for different pull instances, whereas the content of an image for the same digest will always be the same. Pulling images using their digest will give you the confidence that they will always have the same bits inside.

- **Pulling images only from a certain image repository** – If there is a requirement that a container can only pull images from an internally hosted container registry, this rule can help.

- **Allowing images with a specific prefix or patterns using wildcard characters** – If you only want to permit images to be pulled for a specific project, then you can set up a criterion such as allowing images for myapp/*.

- **Allowing images with specific tag values** – This is like the previous rule but applicable to tag names.

In this section, we will learn how to configure a policy that requires image pulls only with digest and not with tags. The following are the steps to do so:

1. Go to the **Policies | Assignment** menu from the left-hand navigation bar and select the **Image registry** tab. From there, select the Workspace we have previously created and click on the **CREATE IMAGE REGISTRY POLICY** button as highlighted in the following screenshot:

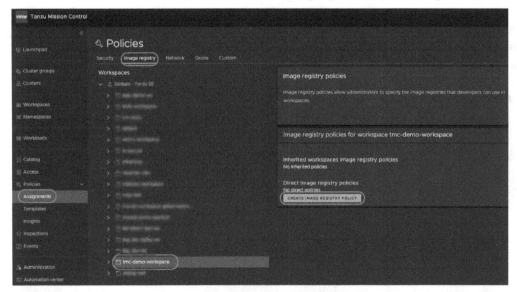

Figure 9.79 – Creating an image registry policy

2. Select the **Require Digest** option from the **Image registry template** dropdown, provide a policy name, and click on the **CREATE POLICY** button:

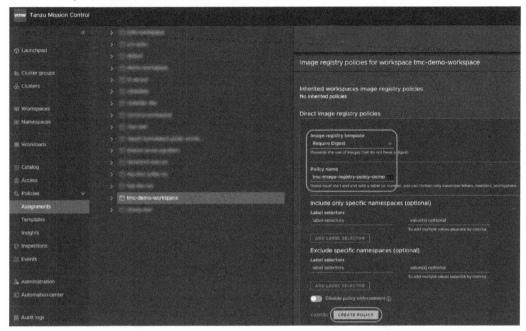

Figure 9.80 – Entering image registry policy details

3. You can see the new image registry policy created for the selected Workspace as shown in the following screenshot:

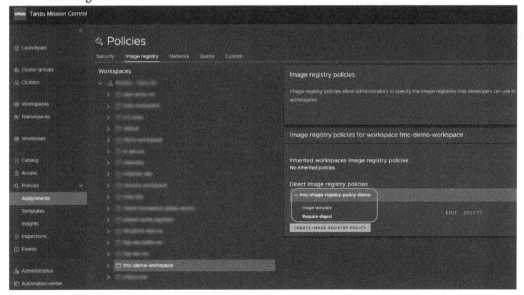

Figure 9.81 – Verifying the creation of the image registry policy

4. Create a Pod in the GKE-attached cluster in a new namespace that is not affected by the policy. This Pod pulls an image using a tag. The Pod definition YAML file is given at `https://raw.githubusercontent.com/PacktPublishing/DevSecOps-in-Practice-with VMware-Tanzu/main/chapter-10/image-tag-using-pod.yaml`:

    I. Create a new namespace in the cluster using the following command:

    ```
    $ kubectl create namespace test-registry-policy
    ```

    II. Create the Pod that uses a tagged image pull:

    ```
    $ kubectl apply -f https://raw.githubusercontent.com/
    PacktPublishing/DevSecOps-in-Practice-with-VMware-
    Tanzu/main/chapter-10/image-tag-using-pod.yaml -n test-
    registry-policy
    ```

5. You will see that the Pod is being created without any issues:

    ```
    NAME                   READY   STATUS    RESTARTS   AGE
    image-tag-using-pod    1/1     Running   0          7s
    ```

6. Create the same Pod in the **default** namespace where we have applied the image registry policy that does not allow you to pull an image with a tag value using the following command:

    ```
    $ kubectl apply -f https://raw.githubusercontent.com/
    PacktPublishing/DevSecOps-in-Practice-with-VMware-Tanzu/
    main/chapter-10/image-tag-using-pod.yaml
    ```

7. You will see that the Pod cannot be created, with the following error message explaining the restrictions in place:

    ```
    Error from server ([tmc.wsp.default.tmc-image-registry-
    policy-demo] container <busybox> has an invalid image
    reference <busybox:stable-uclibc>. allowed image patterns
    are: {hostname: [], image name: [] and require digest}):
    error when creating "image-tag-useing-pod.yaml":
    admission webhook "validation.gatekeeper.sh" denied the
    request: [tmc.wsp.default.tmc-image-registry-policy-
    demo] container <busybox> has an invalid image reference
    <busybox:stable-uclibc>. allowed image patterns are:
    {hostname: [], image name: [] and require digest}
    ```

8. Create the same Pod with a digest replacing the tag for the image using the following command:

```
$ kubectl apply -f https://raw.githubusercontent.com/
PacktPublishing/DevSecOps-in-Practice-with-VMware-Tanzu/
main/chapter-10/image-digest-using-pod.yaml
```

9. You will see that the Pod is created successfully this time in the **default** namespace where the image policy restrictions are applicable:

```
NAME                      READY   STATUS    RESTARTS   AGE
image-digest-using-pod    1/1     Running   0          3s
```

This concludes the topic of learning how to create a policy that is applicable to a Workspace. The previously configured and tested restrictions will also be applicable in the other cluster's **default** namespace, being part of the same Workspace.

In the next topic, we will learn how to apply a deployment governance policy for a cluster group using TMC.

Configuring a deployment governance policy for a cluster group

In the majority of Kubernetes platform deployments, a platform team is responsible for ensuring that their internal customers, the application teams, use the platform with discipline. This discipline involves the fair usage of the computes available and high-availability-prone application deployments. When multiple different Kubernetes platform teams are managing a smaller number of clusters, it will be challenging to implement a set of standards that are applicable enterprise-wide, as all the platform teams may have their own likes and dislikes. On the other hand, if a central management team manages all the clusters, it would be too much to ensure compliance with the governing policies. To address these challenges, TMC allows you to create deployment governance policies. TMC uses an open source project named **Open Policy Agent (OPA) Gatekeeper** (https://github.com/open-policy-agent/gatekeeper) to implement these policies. Because of the declarative nature of the Gatekeeper policy configuration, TMC also allows you to create custom deployment policies for any use case – the sky is the limit!

The following are some of the out-of-the-box policies that TMC provides that can be applied to different cluster groups as required:

*   Blocking the creation of certain Kubernetes resources
*   Mandating the assignment of labels to certain Kubernetes resources
*   Blocking specific subjects from being used for role binding in clusters
*   Restricting the usage of certain load balancer IP addresses to be used by Kubernetes services
*   Enforcing HTTPS for the ingress resource configuration in clusters
*   Blocking the creation of NodePort-type services in clusters

Let's learn how to create one such governing policy for the cluster group that we created earlier and test its impact on the cluster operations. In the following example, we will create a policy to prevent the creation of a Pod without a label named *app*. In other words, creating a Pod in the cluster, where this policy is applied, should fail if that Pod does not specify the name of the application it belongs to.

Take the following steps to create and test this policy:

1. Go to the **Policies | Assignments** menu from the left-hand navigation bar and open the **Custom** tab. Then, select the cluster group we created from the list and click on the **CREATE CUSTOM POLICY** button as highlighted:

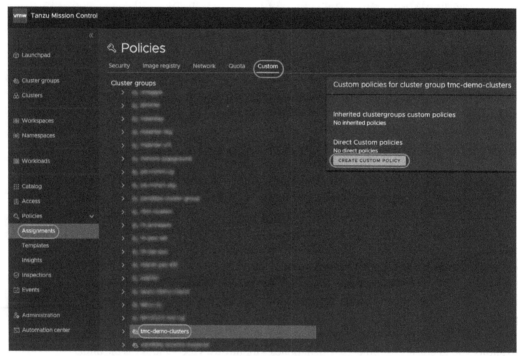

Figure 9.82 – Creating a custom deployment policy

2. Select **tmc-require-labels** from the **Custom policy** dropdown, provide a name for the policy, select **Pod** as the target resource to which to apply the policy, and add `app` as a required key under **Labels**. Finally, scroll down and create the policy:

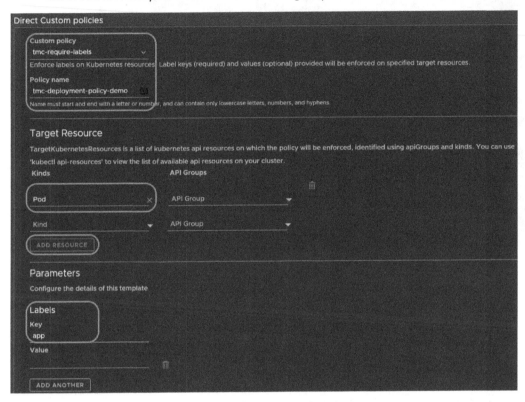

Figure 9.83 – Entering the custom deployment policy details

3. After creating the policy, you should be able to see these settings configured under the cluster group as shown in the following screenshot:

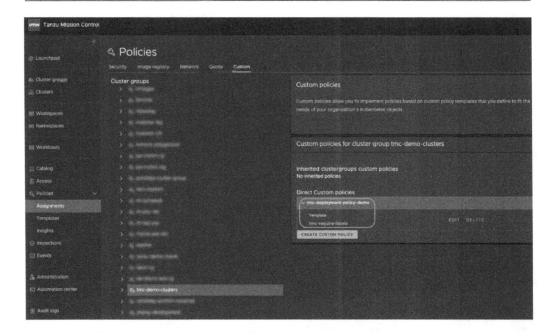

Figure 9.84 – Verifying the creation of the policy

4. Create a Pod in one of the two clusters (the TKG workload or GKE) that we added to the cluster group with the policy applied using the following command. You can see the Pod definition file used in the following command and verify that the Pod does not have any labels in the specification YAML:

```
$ kubectl apply -f https://raw.githubusercontent.com/
PacktPublishing/DevSecOps-in-Practice-with-VMware-Tanzu/
main/chapter-10/pod-without-label.yaml
```

5. You should see a similar error message as follows:

```
Error from server ([tmc.cgp.tmc-deployment-policy-
demo] You must provide labels with keys: {"app"}):
error when creating "https://raw.githubusercontent.
com/PacktPublishing/DevSecOps-in-Practice-with-VMware-
Tanzu/main/chapter-10/pod-without-label.yaml": admission
webhook "validation.gatekeeper.sh" denied the request:
[tmc.cgp.tmc-deployment-policy-demo] You must provide
labels with keys: {"app"}
```

6. Now, let's create a Pod with the required label using the following command. This time, we will use a different Pod specification file where the app label is specified. Open the file used in the command to see the newly added label – `app`:

```
$ kubectl apply -f https://raw.githubusercontent.com/
PacktPublishing/DevSecOps-in-Practice-with-VMware-Tanzu/
main/chapter-10/pod-with-label.yaml
```

7. You will see that, this time, the Pod is created without any issues, as we supplied the label as required by the policy we created earlier in this section:

```
$ kubectl get pod
NAME             READY    STATUS     RESTARTS    AGE
pod-with-label   1/1      Running    0           117s
```

This concludes how to create a custom policy that is applicable to all the clusters in a group and test its impact. TMC admins can create these policies for any logical requirements and create their templates. Later, these templates can be used with custom parameters (such as the name of the label key in the previous example) to apply the policy for a group of clusters – but how do we keep track of policy compliance failures to stay fully informed about these issues popping across hundreds of clusters managed by TMC? Let's find that out in the following section.

Checking policy violation statuses across all clusters

As discussed previously in this chapter, there are several different types of policies we can create using TMC at the cluster group and Workspace level. In addition to creating guardrails for later Kubernetes platforms, TMC also provides a way to monitor them for policy violation insights across all the clusters under TMC's purview. To do so, just open **Policies** > the **Insights** menu from the left-hand navigation bar and you will see a detailed report, as shown in the following screenshot. As you can see, it shows a couple of records for the policy testing we did that violated our policies:

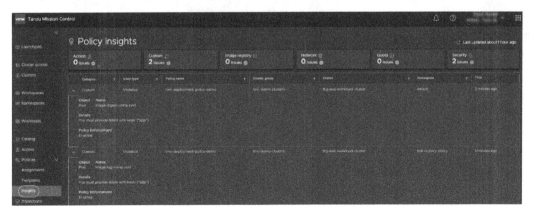

Figure 9.85 – Getting policy insights

Having learned about policies and checking their compliance status, let's now learn how to audit clusters for CIS benchmarking.

Inspecting clusters for CIS benchmark compliance

As discussed earlier in this chapter, a security compliance audit of the clusters for common loopholes is a proactive stance to prevent cyber-attacks. Scanning Kubernetes clusters for a set of best security practices with a long checklist is a challenging task unless there is sophisticated automation built to do so. TMC provides this capability for all its managed clusters using an open source tool, Sonobuoy, as discussed earlier. As of the current version, TMC can perform the following two types of cluster inspections:

- *Conformance* – to check the cluster configuration against the official Kubernetes specification
- *CIS benchmark* – to check whether the Kubernetes cluster deployment is following the security best practices outlined in the CIS benchmark for Kubernetes

Let's learn how to inspect a cluster managed by TMC for CIS Benchmark. Take the following steps to perform the inspection and check the inspection results using the TMC console:

1.   Click on the **Inspections** menu in the left-hand navigation bar and click on the **RUN INSPECTION** button as highlighted in the screenshot:

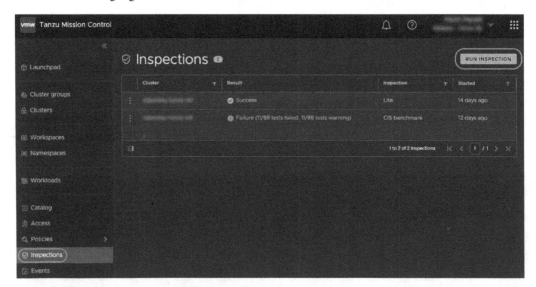

Figure 9.86 – Running a cluster inspection

2.   Select an existing cluster, set the inspection type as **CIS benchmark**, and click on the **RUN** button:

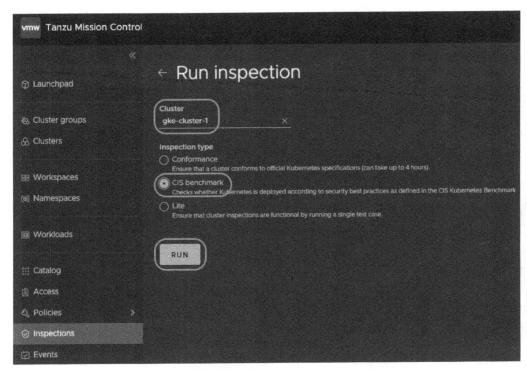

Figure 9.87 – Selecting the inspection to run

3. You will see the inspection results in a few minutes as displayed in the following screenshot. As you can see, the scan has found some failed test cases that should be addressed for a stronger security posture:

Figure 9.88 – Checking the inspection results

4. Click on the test result link on the previous screen and you will get a detailed report of the failed test cases, warnings, and successful cases, as shown in the following screenshot. The report provides details of each test case and recommendations for it. You can also download the report to look at it offline:

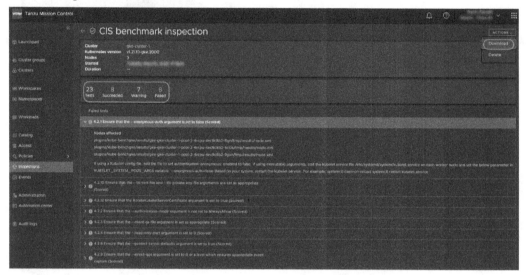

Figure 9.89 – Checking the detailed inspection report

That brings us to the end of the topic of running cluster inspections. We can also trigger cluster inspections from the cluster detail sections as well on TMC in addition to the main **Inspections** menu.

With that, we have also covered all the planned operations that we wanted to cover around applying governance policies for Kubernetes clusters using TMC. First, we saw how to create a security policy to restrict the ability to run privileged containers. Then, we learned how to create an image registry policy, with the example of preventing image pulls without its digest. After that, we created a deployment policy to prevent Pods from being created without a specific label and saw all the different types of custom policies we can create using TMC. Later, we saw how to find policy violations. Finally, we learned how to run conformance inspections for clusters and get their detailed reports. In addition to all these capabilities of TMC, there are a few other operations that we did not cover in this chapter:

- Managing user access policies for cluster groups, clusters, Workspaces, and namespaces

- Installing various tools from a published catalog on a cluster managed by TMC

- Inspecting various events emitted by the clusters managed by TMC

- Performing several administrative activities for TMC, including configuring TMC access permissions, external integrations, and proxy configuration, among other things

To get more details about them, refer to this documentation link: `https://docs.vmware.com/en/VMware-Tanzu-Mission-Control/index.html`.

Summary

We covered a lot of ground in this chapter considering what TMC can do to minimize the effort of managing any distribution of upstream Kubernetes that could be running on different cloud environments. The high-level value proposition of this tool is to provide management control for multi-cloud, multi-cluster, and multi-team usage. First, we understood the different challenges of deployment approaches with a small number of large clusters or a large number of small clusters. Then, we saw use cases for TMC and the solutions it provides for complex problems when managing Kubernetes deployments. We discussed how challenging it can be to operate on different clusters deployed in different cloud environments and keep the operator and developer experience consistent across them. We saw how TMC makes this easy using cluster groups and Workspaces, a group of Kubernetes namespaces across different clusters.

Then, we learned how to get started with TMC, covering the way to integrate a TKG management cluster with TMC. Then, we discussed how to perform cluster lifecycle operations on that TKG foundation using TMC. After that, we saw how to create new workload clusters under the TKG control plane, which become a part of the TMC-managed clusters. Then, we also learned how to bring externally created or managed clusters such as GKE clusters into TMC and make them part of the cluster groups defined in TMC. We also learned how to create groups of Kubernetes namespaces, the Workspaces, for clusters that are deployed in two different cloud environments. After getting started with TMC, the first thing we learned was how to protect cluster data with the backup and restore capabilities of TMC. Then, we covered a long section on governing a fleet of clusters using various policies that we can apply either at the cluster group or Workspace level. Lastly, we covered how to inspect different clusters to take a proactive stance toward security.

As TMC is a single pane of glass for *controlling* large-scale Kubernetes deployments, VMware Aria operations for Applications is a single pane of glass for *observing* the scale of a Kubernetes deployment with its full-stack observability capabilities, starting from the application layer and moving to the infrastructure layer. In the next chapter, we will cover this topic in detail.

10
Realizing Full-Stack Visibility with VMware Aria Operations for Applications

In the last chapter, we learned how we can use Tanzu Mission Control to manage hundreds of Kubernetes clusters for their lifecycle management, policy control, and data protection. Tanzu Mission Control is a single point of control for all these concerns for Kubernetes. In this chapter, we will talk about a central point of visibility for your cloud-native apps running in containers, along with all other supporting layers and systems around them. Knowing the vital health status of your systems is a very old and essential concept in the world of information technology that is known as *monitoring*. It is also known as telemetry. The tools we used for years to monitor traditional monolithic applications would not yield the required results when it comes to monitoring microservices running on different clusters, platforms, and clouds. With microservices, the world is very distributed. Typically, microservices run in containers, and containers are ephemeral – they quickly come and go. In that scenario, the continuity of a microservice's context becomes very important, as they could be running in a different container after a few minutes, unlike traditional apps that would not leave their virtual and physical host for years. Adding more to this complexity of running microservices, they have to deal with substantial cardinality, including the larger apps they belong to, the environments they are deployed to, and the node, cluster, availability zone, and regions they are deployed to, among other things. We need a different approach to monitor them, as they run differently. To some extent, deploying apps in containers is like sending a spacecraft on a Mars mission. We don't know where the container will physically end up – we can only rely on the health vitals that it emits at a regular frequency and course-correct if required. Traditional monitoring tools would not have helped in these conditions. When we have millions of transactions, thousands of containers, hundreds of nodes, and tens of data centers, we need powerful observability. VMware Aria Operations for Applications (formerly known as Tanzu Observability by Wavefront) is the tool in the Tanzu portfolio that addresses this need for modern application management.

In May 2017, VMware acquired Wavefront, a privately held company in Palo Alto, California. The power of Wavefront comes from its ability to ingest millions of metrics and other data points coming from hundreds of locations in real time and render point-in-time charts and alerts for correlated visibility. Later, Wavefront became a part of the Tanzu portfolio to add an essential piece of the puzzle to the mix – observability. Hence, any reference to Wavefront in this chapter or the standard product documentation refers to VMware Aria Operations for Applications. For brevity, we will refer to it as *Aria* in this chapter and will cover it in detail with the following topics:

- **Why Aria?** – covering various features and capabilities of this tool
- **Unboxing Aria** – covering key concepts, components, and the common deployment architecture of the tool
- **Getting started with Aria** – covering the integration of a Kubernetes cluster and an application with the tool
- **Working with charts and dashboards** – covering a high-level understanding of building charts and creating new dashboards with the tool
- **Working with alerts** – covering the details of how to create, manage, and observe alerts with the tool

There are several other facets of Aria that we have to skip to keep the chapter at an acceptable length. Aria is a very powerful tool that can perform various operations around observability and **application performance management** (**APM**), but considering the broader scope of the book, we will only cover the details related to microservices, containers, and Kubernetes monitoring. With that understanding, let's begin our journey of observability.

Why Aria?

Aria is an observability tool rather than a monitoring tool. A monitoring tool can tell you that an application is running slowly, but an observability tool can help you find the root cause for the application being slow. This is because it allows you to correlate the health indicators coming from all the surrounding components that could impact an application's performance. It could be an issue at the **operating system** (**OS**) layer or a slow-running query in a database. The main strength of an observability tool is its ability to ingest data points from all possible systems and layers and for all different health indicators that could make a significant event. It then allows you to find the needle in the haystack by reducing the noise of irrelevant data by applying correlation formulas to the collected data. An observability tool can help you identify abnormal traffic patterns, latencies, error rates, and many more attributes, based on historical data patterns. For example, the average request rate per second for an application during the midnight hours would be different from that during the day generally. Hence, an observability tool could alert you if it saw an abnormal request rate for an application depending on the timeframe. The abilities of these tools help identify anomalies long before they become too-late, expensive discoveries.

With the rise of microservices and containerized platforms, *observability* has become a buzzword in recent times. There are various open source and commercial solutions available on the market offering somewhat similar capabilities in this space. Like other chapters, the idea of this section is not to compare Aria with any other observability tool, but to highlight what makes Aria a compelling choice with its own unique capabilities. Let's look at some of the points in this regard to answer *Why Aria?*

Integrating (almost) anything

As discussed before, the power of observability depends on the data being ingested from various layers and the tools supporting critical business systems. For this, we may need to ingest metrics, events, histograms, and logs from many different systems. They could be various infrastructure platforms such as **Amazon Web Services** (**AWS**) or Azure public clouds; on-premises vSphere stacks; OSs such as Windows, Ubuntu, Photon, or RedHat; middleware layers such as the Tomcat server; Spring Boot- and NodeJS-like application frameworks; various caches; RDBMSes; NoSQL and queues as data sources; and several other components for alerting, containerization, visualization, end user analytics, and so on. To cater to this need, Aria supports over 250 integrations out of the box. These integrations are well documented to aid self-help configuration. Many popular integrations such as Kubernetes, popular public cloud services, and application frameworks come with canned sets of dashboards and alerts, facilitating the quick value realization of these integrations. With these out-of-the-box dashboards and alerts, users can see meaningful data within a few minutes of it starting to flow in. The following are the categories of the out-of-the-box integrations with Aria:

- VMware products
- Web application platforms such as .NET, Tomcat, and nginx
- Cloud services of AWS, Azure, and **Google Cloud Platform** (**GCP**)
- Data stores such as PostgreSQL, Redis, Cassandra, and Oracle
- DevOps tools such as Jenkins, GitHub, Chef, Ansible, and Terraform
- Messaging platforms such as RabbitMQ and Kafka
- Monitoring tools such as AppDynamics, Dynatrace, and Prometheus
- OSs such as different flavors of Linux, Windows, and macOS
- Application frameworks such as Spring Boot, Python, Go, and Java
- Alerting systems such as PagerDuty, ServiceNow, and Slack
- Authentication providers such as Okta, Google, and Microsoft Active Directory

This list is not comprehensive. There are various tools in each category beyond those listed here. On top of that, Aria has an extendable framework that allows you to pull in metrics and other health indicators from any system using its plugin-based model and a **software development kit** (**SDK**). As we will see later in the chapter, Aria can get metrics from most systems using an open source metrics collection agent named **Telegraf** (`https://github.com/influxdata/telegraf`). Telegraf has a long list of plugins that make it very extendable. For the data sources not available as integrations in Aria out of the box, we can create custom integrations with minimal effort. Additionally, in case an out-of-the-box plugin is not available to ingest data from the source, it can be custom-developed using the Aria SDK. Visit `https://docs.wavefront.com/wavefront_sdks.html` for more details.

Getting full-stack visibility

Have you ever realized that when you ran an application in containers, how many different layers were below and around that application? An application is wrapped in a container. In Kubernetes, one or more containers are wrapped in a Pod. These Pods are a part of a Kubernetes node. A node is often a virtual machine that is running on top of a hypervisor such as vSphere. The hypervisor sits on top of physical hosts. These physical hosts are parts of some rack in a data center somewhere in the world. Additionally, the application could be using external services, such as other dependency apps, an application runtime such as the Tomcat server or Java Runtime Environment, a database, possibly a cache, a messaging queue such as RabbitMQ or Kafka, and many other supporting services. When we need to monitor an environment of many containers, monitoring all these layers below applications and their supporting services becomes crucial to quickly find out the root cause of failures and quickly address them. This is made possible when you have the ability to collect health vitals from all these sources and visualize them meaningfully or establish an alerting system using a single tool. **Site reliability engineering** (**SRE**) teams often struggle when there are different tools for monitoring different systems, such as a separate tool for application monitoring, a separate tool for Kubernetes monitoring, a separate tool for virtual infrastructure monitoring, and so on. When we have several monitoring tools to investigate during an outage situation, making sense of the health data available becomes very difficult and results in a loss of context. This happens because different tools use different methods of rendering the collected health data and the teams using them use different terms and language. In these conditions, a common way to filter the metrics collected for a specific timeframe and other conditions for all the sources potentially affected can quickly help corner an issue. This way, we can quickly point out that a slow-running application in a container is actually being slowed down by a high CPU host temperature condition.

Since Aria can collect health data from almost any source, we can create custom dashboards for custom applications and display health vitals from all the surrounding components related to the application. This way, you can see the memory utilization of an application's container and of the Kubernetes node on which the container is deployed.

Ingesting high-volume data in real time

Aria is a very powerful cloud-based SaaS streaming analytics platform that is highly scalable and efficient at collecting a very large amount of data. Because of its design, it can potentially support collecting over a million data points per second. Once this data is collected, we can see it live in its respective monitoring dashboards and it can be used to calculate any preconfigured alert conditions with Aria's powerful query engine to pull ingested data in real time.

Retaining full-fidelity data for a long time

As we discussed previously, Aria can ingest a very large volume of data – but additionally, it can also retain all metrics data for 18 months as of the time of writing. This is unlike other observability tools that either store this data for a shorter time period or with some level of aggregation applied to it to reduce the amount after a specific period. With the help of Aria's full-fidelity metrics retention capability, we can compare a system's performance and state for a past timeframe as long as 18 months. However, health data types such as histograms and span logs have a smaller retention period. We will discuss histograms and span logs later in the chapter.

Writing powerful data extraction queries

Ingesting a lot of data from various sources with a long retention period in a time-series database is of little use if we cannot pull it in the way we want with specific filters, aggregations, and correlations to other similar data points. Aria has a detailed query language that enables data extraction with all these abilities. These queries can incorporate a subset form of regular expressions (Regex), wildcard characters, aliases for simplified references, variables, relational operators, arithmetic operators, and several types of data manipulation functions. These functions can be categorized as follows:

- Aggregation functions such as sum, average, minimum, maximum, and many others

- Filtering and comparison functions such as between, top, bottom, random, and many more

- Time operation functions such as rate, rate difference, year, month, day, and many more

- Moving window functions for aggregated operations for a moving time window of data – for example, getting the average CPU usage of a host for the past hour

- Missing data functions to replace missing data values with specific values

- Conditional functions such as `if` blocks

- Exponential and trigonometric functions such as getting square roots, exponential, sin, cos, and a few more

- Metadata functions temporarily renaming metrics and sources or creating a custom point tag on the time series of data values

- String functions to manipulate string values

- Predictive analytical functions to predict certain values or find outliers

- Histogram processing functions to manipulate ingested event data

- Event processing functions to manipulate ingested event data

- Distributed traces and spans functions to find and filter trace data sent by applications

- **Application performance index (Apdex)** score functions

> **What is an Apdex score?**
>
> Apdex is an open standard intended to simplify reports on application performance. Apdex analyzes the perceived satisfaction of the application's end user. It is not an APM tool. Apdex numerically scores the level of satisfaction of an end user based on the application responsiveness by calculating the degree to which user expectations compare to the performance on a fractional 0 (no users satisfied) to 1 (all users satisfied) scale. *Source: TechTarget.com*

Aria supports around 200 different functions cumulatively under these categories as a part of its query language, which provides a great level of flexibility to get what we want.

Getting SaaS benefits

Since Aria is a SaaS offering, we can quickly get started without heavy preparation. With a few steps of integration, we can start getting the value out of it – and like most other SaaS solutions, it also supports a *pay-as-you-go* billing model calculated in terms of **points per second** (**PPS**). To understand what PPS is, let's review an example. Let's say there are 6 containerized apps and each of them sends 10 different metrics every 60 seconds with a total of 60 metric data points per 60 seconds from both containers. The Kubernetes cluster where these containers are deployed sends 50 different metrics at a frequency of every 10 seconds. That means we get 300 metric data points total ingested in 60 seconds for the cluster. Therefore, cumulatively, we have 360 data points from the cluster and the containers running on them in 60 seconds, so that means the data ingestion rate is 6 PPS (360/60). In this way, Aria calculates the total PPS applicable for the entire account, which is used for billing purposes.

With this, we have concluded our section explaining the potential reasons that make Aria a valuable tool in your toolkit when it comes to managing modern applications, especially running in containers. We saw how Aria helps you get full-stack visibility by bringing in various sources that can impact your applications. We also learned about its capacity to ingest a huge amount of data, retain it for a long time, and extract it with a comprehensive query language. In the next section, we will learn more about Aria including some concepts, terms, and its high-level deployment architecture.

Unboxing Aria

After understanding the reasons why and the use cases for which Aria can be a valuable tool to comprehensively monitor your cloud-native applications, let's now learn more about the tool to understand what different data formats it can capture, its deployment architecture, and the building blocks that comprise this distributed system.

Supported data formats in Aria

The following is the list of different data formats that Aria can ingest and then use to generate useful charts, dashboards, and alerts.

Metrics

A metric is a small text-based record that carries the state data of a source being monitored at a specific timestamp. It may optionally have other tags for additional metadata that can be used to build useful queries for joins and filters. The most common form of metrics is time-series-based. We use it to report things such as the amount of a server's memory occupied at any given point in time:

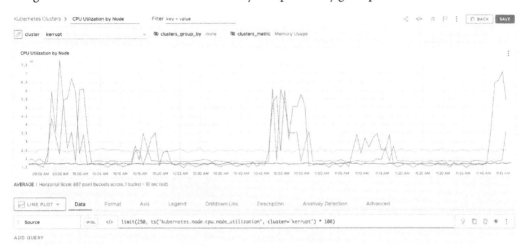

Figure 10.1 – A time-series metrics chart example

Figure 10.1 gives us a simple chart in Aria showing how time-series metrics data would look. In this case, the chart shows the CPU utilization of the nodes of a cluster named **kerrupt** for the given period.

A metric can contain a **gauge** value, which is the current value of the indicator for each point in time, such as the current available CPU for a server under observation. The values of the gauge metrics are not related to the previous values. A metric may also be of a **counter** type, which is an incremental value at any point in time, such as the number of orders placed since the counter was reset. Finally, a metric can also be of the **delta counter** type and report a subset value of a group of similar sources. For example, if you have multiple copies of a containerized application that processes orders, each copy of those application containers can report a delta value of the orders that it has processed. In these conditions, there is no way that one container can report the total value of the orders processed by all the copies of that application's containers. In that case, Aria can group these delta counter values to produce a cumulative value of the total orders processed by the application. Learn more about metrics in Aria here: `https://docs.wavefront.com/metrics_managing.html`.

Histograms

While Aria can receive a large number of data points per second as a whole system, it can only store one data point per second for a combination of a source and its metric name. For example, you have a load balancer in front of a very busy system, which receives hundreds of requests per second. In that case, if we send a metric value to be stored in Aria for each request, then Aria would be able to store only one value in its database for that metric and from that source. It can store multiple values in a second only if either the source or the metric name is different. In that case, if we want to report on the response time for each request, this is not possible because we can only store the response time of one request per second for the same source. In these cases, we can use histograms that show the distribution of data in a given period. In our example, we can report a histogram containing the distribution of requests based on their response times for a minute, an hour, or a day. *Figure 10.2* shows how this kind of histogram would look:

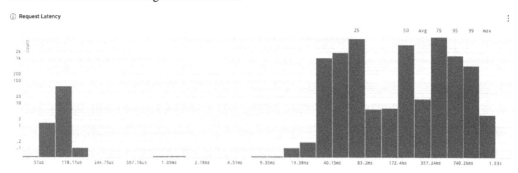

Figure 10.2 – A histogram example

In *Figure 10.2*, on the *y* axis, we have the number of requests falling in one bracket. On the *x* axis, we have the response time in milliseconds for the requests. Seeing this histogram, we can say that most of the requests fell between the range of 40 to 800 milliseconds in their response times. This way, we know what a normal acceptable range for the response time value is for these requests. Next time we see a greater number of requests falling within the higher response time bracket, we will know something is wrong there.

Events

In general terms, an *event* is something interesting that occurred in a system under observation. In the world of information technology, events can be a new application version rollout or an alert condition being triggered. Aria generates events automatically in cases of any alert being triggered. Additionally, it also allows you to manually create event records if required. We can see these events displayed in the applicable charts that are related to the event sources. For example, if we create an event indicating a new version of an application being pushed, we can clearly compare the difference in response time for the newer version compared to the older one, as its corresponding chart would display an event marker with these details of the application rollout. *Figure 10.3* shows how Aria displays these events in applicable charts. As you can see, an event can also be a timeframe that may have a different start and a different end time:

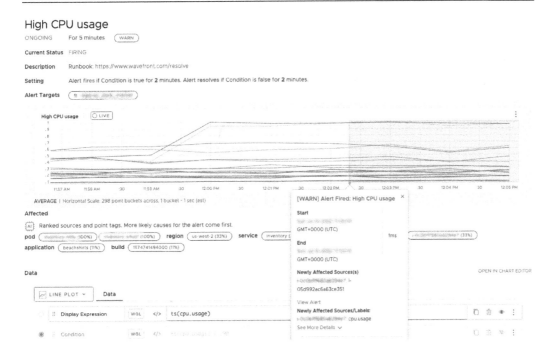

Figure 10.3 – A chart showing events

As you can see, *Figure 10.3* shows an event triggered by a warning level of the alert condition. The shaded part after the alert dot shows the duration when the alert condition is active, and the event was ongoing when the snapshot was taken. Ongoing events of this kind can be closed manually, closed when the alert condition gets resolved, or Aria closes them after 60 days.

Span logs

At a very high level, span logs report the health of the communication channel between any two services in a distributed system. In the world of microservices, we commonly see a single user request traversing various sequential and parallel service calls involving many microservices and third-party systems. For example, in an e-commerce application, an order submission request by an end user could involve calls to a payment service, a payment method verification service, an order provisioning service, an inventory adjustment service, and a database before the end user gets a response. In this case, the request/response channel between any two microservices is called a **span**. **Span logs** can report how much time is taken for the request/response between any two microservices. For example, in this case, span logs can report the request/response duration between the order service and the payment service, the payment service and the payment verification service, and so on. When all these spans form a single request flow, it is called a **trace**. These are all concepts of distributed tracing, a very powerful way to monitor the health of microservices, especially when they are deployed in containers. Learn more about distributed tracing concepts here: `https://docs.wavefront.com/tracing_basics.html`.

In order to provide distributed tracing for applications, Aria supports the **OpenTracing** (https://opentracing.io/) and **OpenTelemetry** (https://opentelemetry.io/) open source standards. However, OpenTracing is now an archived project, and it is recommended to use its better replacement – OpenTelemetry – a **Cloud Native Computing Foundation** (**CNCF**)-governed project. OpenTelemetry provides the tools, APIs, and SDKs required for applications written in different languages to publish span logs. Once applications send their span logs into Aria, we can get application maps like the one shown in *Figure 10.4*:

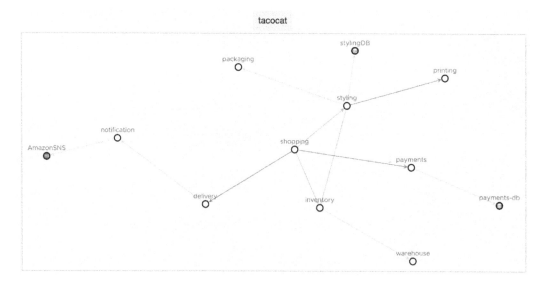

Figure 10.4 – An application map example

Figure 10.4 shows a distributed tracing map for an e-commerce application, where all the circles indicate the different services involved. Additionally, the arrows indicate the flow of requests from one service to another and the color of the arrows indicates the health of the systems in terms of the response time and error rate. Having proper span logs in Aria from all services involved can be very beneficial to uncover hidden performance issues in a matter of minutes. Learn more about other possible visualizations in Aria generated using span logs here: https://docs.wavefront.com/tracing_basics.html.

With these four data types that Aria can ingest, including metrics, histograms, events, and span logs, we can get the required observability for our applications, their supporting systems, and the infrastructure they are running on. Let's now understand a common deployment architecture pattern that allows these data points to be ingested in an Aria database on its SaaS cloud.

Data integration architecture of Aria

As discussed previously, Aria is a SaaS platform supported by two optional components – the collector agents and proxy service. *Figure 10.5* shows a high-level diagram of different ways to ingest data into the system:

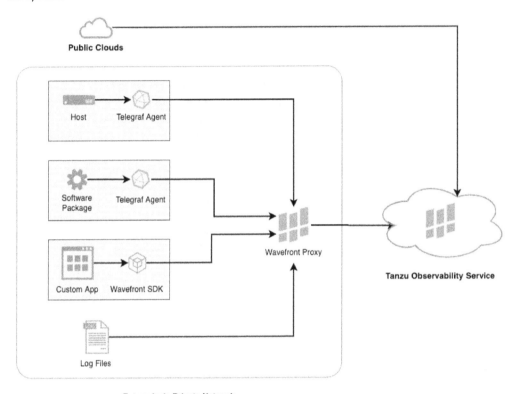

Figure 10.5 – Data integration architecture of Aria

Let's understand the deployment architecture depicted in *Figure 10.5*. This diagram has the following components.

Data source

The large circles shown on the left-hand side of the diagram are the sources from which the health data is collected. There could be the following types of sources, as shown in *Figure 10.6*:

- **Public clouds** – These are the major public cloud providers, including AWS, Azure, and GCP. We can configure the data flow from these cloud services using an account-level configuration on the Aria portal without requiring any collector agents on those clouds.

- **Host** – This could be a hardware device, an OS, a network component, or anything that emits health data points, which we covered earlier. Generally, you cannot configure a customized endpoint for the metrics data with these components. They just *tweet* metrics. We need metrics collector agents for them as well that can then send the data to the Aria database.

- **Software package** running on a host – These can be any software solution that is deployed on a host (server). It could be a MySQL database or a Jenkins server. We can collect their metrics using a collector agent that knows how to collect the data emitted by these software tools. We will revisit these agents later in this chapter.

- **Custom app** codes – These can be any custom business application that sends custom business metrics, such as order counts or failed payments. This data can also include span logs for specific operations emitted by equipped applications. This data can be ingested into Aria from the application code using certain third-party libraries depending on the technology stack of the application. Micrometer (`https://micrometer.io/`) is one such open source library that is widely used for custom metrics emissions from a Java-based application. Similarly, OpenTelemetry (`https://opentelemetry.io/`) provides several libraries for various application technologies to emit metrics and span logs from custom applications.

- **Log files** – There are some systems that do not emit any telemetry data but just write them in local log files. We can also collect this data from log files using specific configurations in Wavefront Proxy. We will cover Wavefront Proxy shortly

These are all different types of sources from which we can collect telemetry data. While some sources can directly send data to the Aria system, for many sources, such as software running on servers such as databases, off-the-shelf solutions, or sources such as a host OS, collector agents need to gather this data and send it downstream. Let's learn about them.

Collector agents

As mentioned before, many sources just emit telemetry data without any knowledge of which system will collect them for processing. These sources often have a source endpoint at which they make this data available for collection. To collect this data from its source, filter it for the subsets not required, aggregate it when required, and convert it into a form that can be understood by a downstream telemetry visualization system such as Aria, we need a tool that can help with these requirements. The agents shown in *Figure 10.6* next to the **Host** and **Software Package** components are there for this very same reason.

Aria is a time-series database for ubiquitous observability that accepts data from different sources in a specific format that it understands. For this reason, Aria requires collector agents for sources that make their telemetry data available at a specific endpoint and require their consumers to pull it as and when required. While we can write custom programs that can perform this task, there is a very elegant out-of-the-box solution available that we can directly use as a collector agent – **Telegraf** (`https://github.com/influxdata/telegraf`).

Telegraf is a mature and flexible open source metrics collector agent tool that has a very extendable plugin-based architecture to support collecting, processing, and distributing telemetry data. Telegraf has the following types of plugins that make it a very popular and widely adopted tool:

- **Input** – These plugins allow you to collect data from different source types. Each source has a specific plugin associated with it that can be used to collect its telemetry data.

- **Processor** – These plugins allow you to modify and filter collected data.

- **Aggregator** – These plugins allow to aggregate data before it is sent to a source to reduce the volume.

- **Output** – These plugins are used to send data to a specific output destination. In our case, it is Wavefront Proxy, which we will cover next.

Aria can accept telemetry data via an HTTP API endpoint. Hence, a data source can always use that to ingest data in Aria. However, for sources that only make their health data available for collection, we need the help of tools such as Telegraf to fill the gap for sure – but rather than sending such data directly to the Aria database, it is recommended to send them via Wavefront Proxy. Let's learn about Wavefront Proxy next.

Wavefront Proxy

Wavefront Proxy is an open source tool (`https://github.com/wavefrontHQ/wavefront-proxy`) written in Java that sits between a collector (and sometimes between the source directly) and the Aria database on the cloud. Although using Wavefront Proxy is not mandatory, it is a very useful component in Aria architecture. As a reason, VMware actively maintains this project. Architecturally, Wavefront is placed closer to the data sources. As shown in *Figure 10.6*, several sources can share a single Wavefront Proxy, which could be deployed on the same host, data center, or private network boundary as the sources. The following are the reasons to consider using Wavefront Proxy, rather than sending telemetry data directly to the Aria database using its API endpoint:

- Wavefront Proxy provides a layer of data protection with the ability to cache data in memory in case the link between the source and Aria data endpoint goes down, which works via the internet.

- Wavefront Proxy batches data for the optimal use of the network bandwidth and an optimized transmission speed.

- Wavefront Proxy provides a way to enrich data by adding more tags (key-value pairs) to each data record it processes. These tags can be used to fetch the details required for monitoring. An example of these tags could be `env=prod`, defining the data record as belonging to a production environment. Later, we can use this tag to build a chart showing only production environment data.

- Wavefront Proxy allows us to massage data to modify the content if required. One such use case could be to hide the specific IP addresses of the sensitive systems to maintain the privacy of the network layout. In general, having the IP address of a source in its telemetry record is a useful detail that can be used to filter data in the same way. However, an organization may need to hide IP addresses as per their policy for data going into an Aria database that resides in a public cloud space. For this, Wavefront Proxy allows us to carry out pattern-based replacement in data records. This way, we can replace an IP address such as `10.1.0.2` with either `10.*.*.2` or `*.*.*.*` or `******`.

- Wavefront Proxy provides the last centralized gate that opens to the Aria service endpoint. If all sources send their telemetry data via the same instance of Wavefront Proxy, we can use that as the final checkpoint to apply common data filters and enrichment policies applicable to all sources.

- Since Wavefront Proxy is the last gate for the internal telemetry data before it reaches the Aria service endpoint, we only need to open the firewall from Wavefront Proxy to the Aria service over the internet. All internal sources can send their data to Wavefront Proxy over the private network. This reduces the network attack surface.

In addition to the preceding benefits, there is one more that is worth a mention here. Very large organizations have several data centers, either in a private or public cloud space. At this scale, it is not advisable to have just one instance of Wavefront Proxy between all the sources deployed across public and private network spaces. As mentioned before, we should try to keep the Wavefront Proxy setup as close as possible to its sources so that we can take advantage of its data reliability and enrichment capabilities. Hence, we generally end up with multiple instances of Wavefront Proxy deployed that send data to the Aria SaaS endpoint. This kind of setup could be difficult to maintain down the line, as we need to maintain configuration parity between these Wavefront Proxy implementations. To ease this situation, we can create Wavefront Proxy chaining, where all distributed instances of Wavefront Proxy send their data to a centralized instance of Wavefront Proxy instead of directly sending the data to the Aria service endpoint. This way we can keep all source-specific configurations in remote Wavefront Proxy instances and all common configurations in the central instance of Wavefront Proxy.

For more details about various deployment patterns and other details of Wavefront Proxy, visit this page: `https://docs.wavefront.com/proxies.html`.

The Wavefront service

In *Figure 10.5*, the right-hand element captioned **Tanzu Observability Service** is nothing but the Aria SaaS platform deployed on the AWS cloud (as of the time of writing). This is also the component that we have been referring to as the Aria service endpoint. The Aria service is made of a large collection of components, including a portal, API endpoints, a time-series database, various data processors, alert engines, and several other components. We will cover working with the Aria portal for various operations in upcoming sections of this chapter.

With this, we have concluded our section on unboxing Aria. In this section, we saw that Aria can ingest metrics, events, histograms, and span logs to provide full-stack observability. Additionally, we learned about the common deployment architecture of Aria. There, we learned that Aria can ingest data from various sources, including public cloud platforms, different hosts and their OSs, and different software running on those hosts, including platforms such as Kubernetes, custom applications, and log files. Then, we learned about the collector agents' role and Telegraf being one of the most popular collector agents. Then, we checked out the role and benefits of using Wavefront Proxy in the data path. Let's now see how to get started with Aria and integrate our first Kubernetes cluster with an Aria service account for monitoring.

Getting started with Aria

In this section, we will learn about the following topics:

- Setting up a trial account for Aria (unless you have an existing account)

- Integrating a Kubernetes cluster with the account for monitoring

- Accessing the default dashboards for the integrated Kubernetes cluster

- Accessing the default alerts for various Kubernetes cluster conditions

Here are the prerequisites to follow along with the outlined steps in this section and upcoming sections:

- A Kubernetes cluster with admin-level `kubectl` access

- A workstation with `kubectl` and the `helm` CLI, access to the targeted Kubernetes cluster, and a web browser

- A valid email address

As you can see, the prerequisites are very straightforward. Let's start by integrating a Kubernetes cluster with an Aria service account.

Setting up a trial account

This section is optional if you already have an existing Aria service account that you would like to use. Otherwise, take the following steps to get started with a trial account of Aria without any obligations:

1. Go to the `https://tanzu.vmware.com/observability` page.

2. Click on the **START FREE TRIAL** button as shown in the following screenshot:

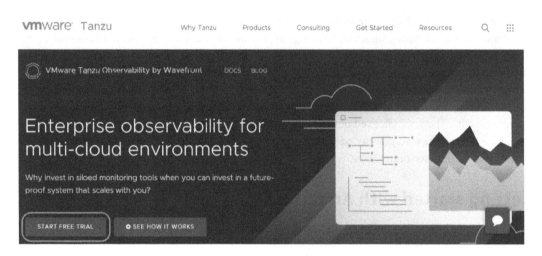

Figure 10.6 – Aria website

3. Submit the trial account setup details as shown in the following screenshot:

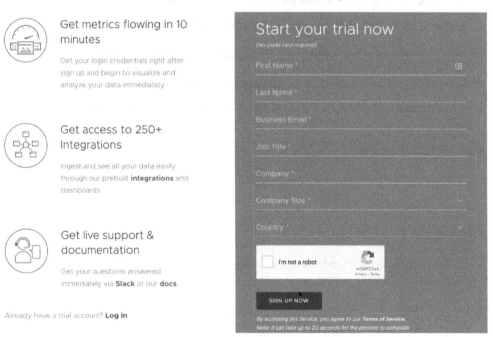

Figure 10.7 – Submitting the trial account setup details

4. Configure a password for the account:

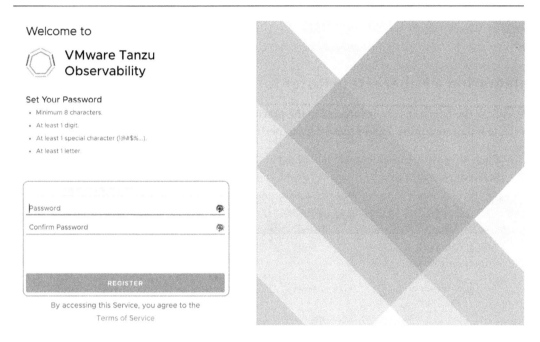

Figure 10.8 – Configuring a trial account password

5. Load the landing page as shown in the following screenshot:

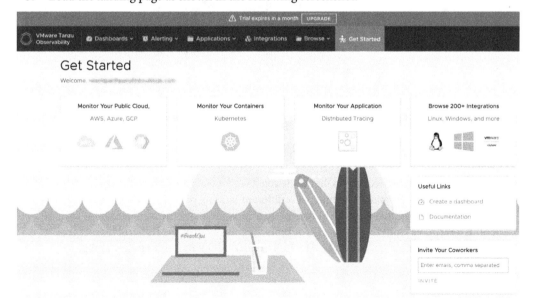

Figure 10.9 – Aria landing page on the first login

That is all. We successfully created a trial account for Aria in a few minutes with all its features accessible. Let's now use this account to integrate a Kubernetes cluster for monitoring.

Integrating a Kubernetes cluster for monitoring

We will use the Aria trial account created in the previous subsection to link an existing Kubernetes cluster to it for monitoring. Follow these steps to perform the integration procedure:

1. Click on the Kubernetes logo as highlighted in the following screenshot from the landing (**Get Started**) page of Aria:

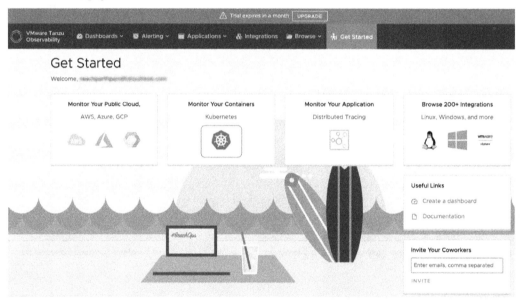

Figure 10.10 – Starting Kubernetes cluster integration

For existing non-trial accounts

You may use your existing account to do this. If you have previously accessed the account, based on your account configuration, you may see a different landing page than what is shown in the previous screenshot. In that case, go to the **Integrations** menu on the top navigation bar and click on **Kubernetes** from the featured group of integrations. On the Kubernetes integration page, go to the **Setup** tab and click the **ADD INTEGRATION** button to follow along.

2. Select the middle tile with the **Install in Kubernetes Cluster** option to load the setup instruction page. The setup procedure for a Tanzu Kubernetes cluster is pretty similar except for a few additional configurations. However, the procedure to integrate an OpenShift cluster is very different, but we are not covering that in this book:

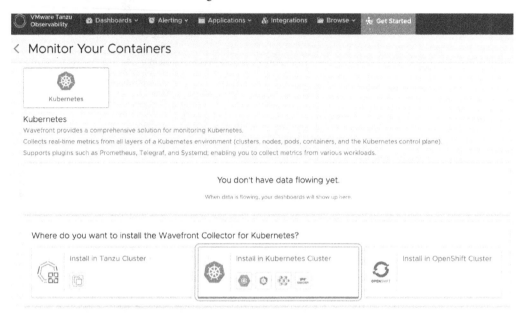

Figure 10.11 – Selecting the Kubernetes type for the integration

3. Follow the set of instructions given on the page to install a deployment of Wavefront Proxy and the metrics collector for Kubernetes provided by Aria. This is a quick and easy way to get started. However, it is always possible to perform various fine-grained configurations provided by the collector as well as Wavefront Proxy for a production-grade integration. Follow this documentation link to learn more about this: `https://docs.wavefront.com/kubernetes.html#kubernetes-manual-install`:

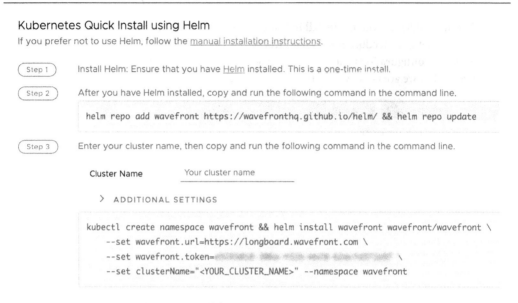

Figure 10.12 – Kubernetes integration instructions

4. After running the command in *step 3*, we can see the following output. Notice the last line providing the URL to examine the cluster integration on the Aria portal:

```
$ kubectl create namespace wavefront && helm install
wavefront wavefront/wavefront\
    --set wavefront.url=https://longboard.wavefront.com \
    --set wavefront.token=**-306a-432b-bb28-62dc**2e97 \
    --set clusterName="eks-workload-cluster-1"
--namespace wavefront
namespace/wavefront created
NAME: wavefront
NAMESPACE: wavefront
STATUS: deployed
REVISION: 1
NOTES:
Wavefront is setup and configured to collect metrics from
your Kubernetes cluster.  You
should see metrics flowing within a few minutes.
You can visit this dashboard in Wavefront to see your
Kubernetes metrics:
https://longboard.wavefront.com/dashboard/integration-
kubernetes-summary
```

5. Check out the deployment of Wavefront Proxy and Wavefront collector Pods in the `wavefront` namespace of the targeted cluster. As you can see, there are two collectors, one for each Kubernetes node, and one Wavefront Proxy Pod:

```
$ kubectl get pods -n wavefront
NAME                                   READY   STATUS    REST
ARTS     AGE
wavefront-collec-
tor-2vbbt               1/1        Running    0           5m51s
wavefront-collec-
tor-xv9sk               1/1        Running    0           5m51s
wavefront-proxy-699f57f698-
1br9b    1/1    Running    0          5m51s
```

6. Examine the Kubernetes integration status using the URL given in the output of the command described in *step 4*. You will see a page as shown in the following screenshot:

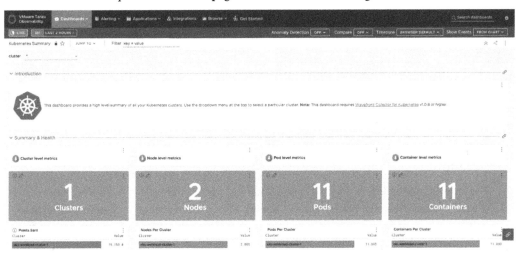

Figure 10.13 – Kubernetes cluster summary dashboard

As you can see, the Aria account has been successfully integrated with an existing Kubernetes cluster. We can add as many clusters as required this way to centrally monitor them from a single window. Then, we can filter a specific cluster if required by its name using the dropdown labeled **cluster,** as shown in the previous screenshot's top-left corner. The dashboard shown in the previous screenshot is one of the many out-of-the-box dashboards in Aria for Kubernetes integration. We can access some of them from this dashboard with a drill-down approach. Let's check out the other out-of-the-box Kubernetes dashboards.

Accessing the default Kubernetes dashboards

In this section, we will access some additional Kubernetes dashboards that come by default with Aria. These dashboards immediately start showing data after a successful Kubernetes cluster integration. Let's start with the first dashboard – **Kubernetes Summary** – that we opened in the previous section. From there, we will access various other dashboards:

1. To drill down to the **Kubernetes Cluster** dashboard from the **Kubernetes Summary** dashboard, click on the cluster name as highlighted in the following screenshot's bottom-left corner:

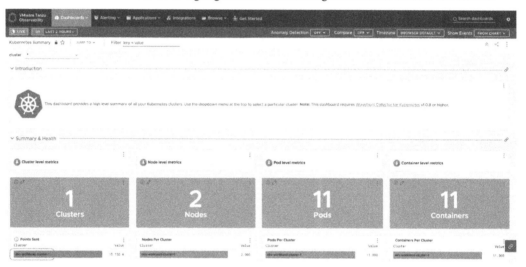

Figure 10.14 – Drilling down to the Kubernetes cluster dashboard

2. You should see the following dashboard showing a more detailed view of the Kubernetes cluster. Then, click on the **Nodes** link as highlighted in the screenshot to open the **Kubernetes Nodes** dashboard for a detailed view of the nodes of that cluster. Additionally, you can find very valuable insights on all these dashboards when you scroll down their pages:

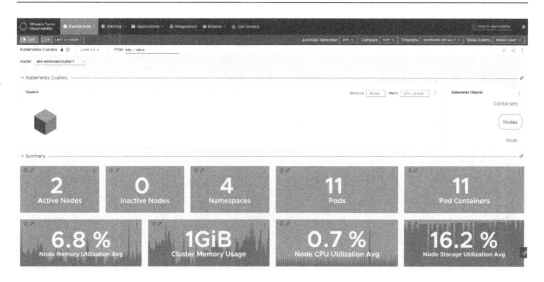

Figure 10.15 – Kubernetes Clusters dashboard

3.   The following screenshot shows the **Kubernetes Nodes** dashboard, which contains a detailed view of the nodes of the cluster to which we are drilling down. You can select a specific node of the cluster using the **node** dropdown as highlighted in the following screenshot. Additionally, you may open the **Kubernetes Pods** dashboard by clicking on the **Pods** link as highlighted:

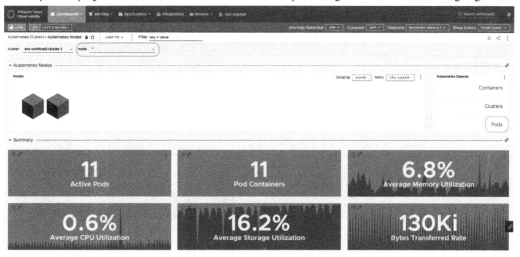

Figure 10.16 – Kubernetes nodes dashboard

4.   Upon clicking on the **Pods** link on the **Kubernetes Nodes** dashboard, we get to the **Kubernetes Pods** dashboard, as shown in the following screenshot. As also shown in the screenshot, you can group these Pods by several criteria. These groupings are very useful when we have hundreds of Pods running in a cluster. Additionally, we can also drill down to the **Kubernetes Containers** dashboard from this page as we have seen in the previous points:

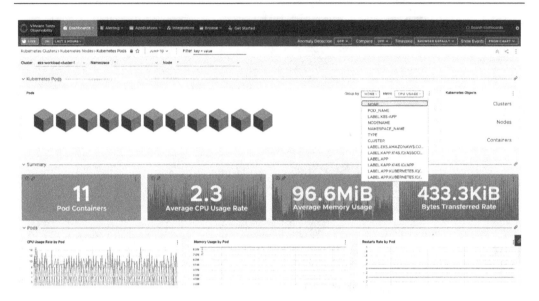

Figure 10.17 – Kubernetes Pods dashboard

In addition to all these dashboards, there are several out-of-the-box Kubernetes dashboards we can access using the following steps:

1. Click on the **Integrations** menu of the top navigation bar as shown in the following screenshot:

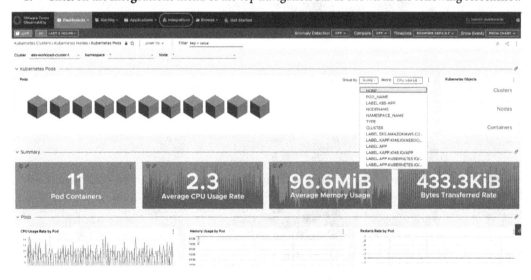

Figure 10.18 – Opening the Integrations page

2. Click on the **Kubernetes** tile as shown in the following screenshot:

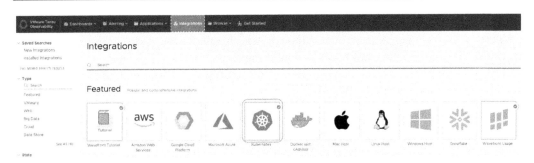

Figure 10.19 – Opening the Kubernetes integration tile

3. Click on the **Dashboards** tab as highlighted in the following screenshot:

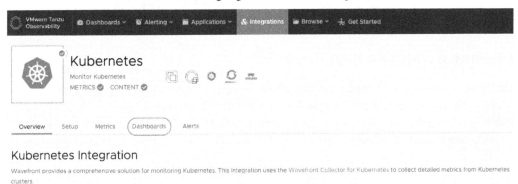

Figure 10.20 – Opening the Kubernetes dashboard page

4. The following screenshot shows all the default dashboards in Aria for Kubernetes monitoring:

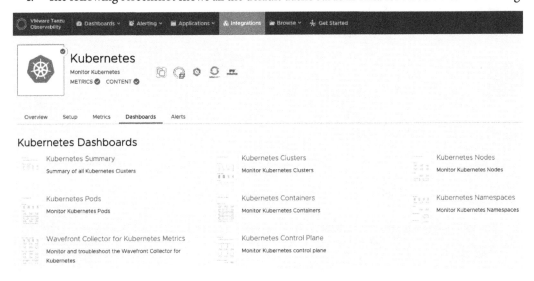

Figure 10.21 – Default Kubernetes dashboards

All these out-of-the-box dashboards show useful insights using the metrics collected from all integrated Kubernetes clusters with an Aria account, although Aria also allows you to clone these dashboards and customize them to address any specific monitoring requirements. We can add new charts and new drill-downs, modify existing visualizations, rearrange different tiles, and make several other customizations to these default dashboards.

However, dashboards are useful only when someone explicitly and carefully observes them to find abnormalities. People generally open them to find anomalies after an issue is reported. This is a reactive analysis of an issue that has already occurred, but we use these sophisticated tools to catch issues well in advance, in a proactive manner, before our customers feel the pain. For that reason, we have to configure alerts in these tools that notify us as soon as any abnormality in any health condition is observed by the monitoring system. Aria has a powerful alerting capability for the very same reason – and like Aria has a default set of ready-to-use dashboards for Kubernetes, it also comes with a set of predefined alert configurations for proactive Kubernetes monitoring. Let's learn about them.

Accessing default Kubernetes alerts

In the previous section, we explored the default dashboards that come with Aria for Kubernetes integration. The same is the case for a set of useful alerts that are supplied with Aria for Kubernetes. These alerts are configured with opinionated criteria. Aria allows you to customize them for various reasons. The following steps show how to access and configure the out-of-the-box alerts for Kubernetes cluster monitoring:

1.   Click on the **Alerts** tab as shown in the following screenshot:

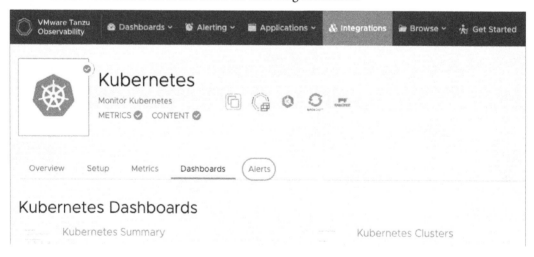

Figure 10.22 – Opening the Kubernetes Alerts page

2.   The following page shows all the default alerts that come with Aria for Kubernetes:

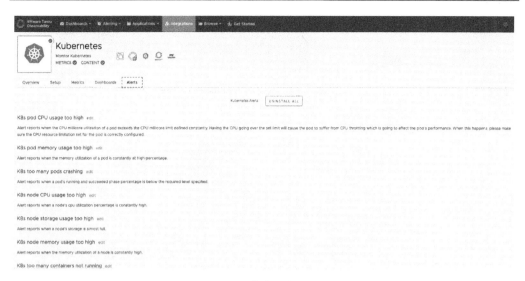

Figure 10.23 – Default Kubernetes alerts

Like dashboards, all these alerts are fully customizable to change their firing conditions and delivery mediums, such as texts, emails, PagerDuty, ServiceNow, Slack, and many others. We will cover more about working with alerts later in the chapter. For now, let's see how to get more details about existing alerts and their status to see whether there are any active conditions. For that, perform the following steps:

1. Click on the **All Alerts** option under the **Alerting** menu from the top navigation bar as shown in the following screenshot:

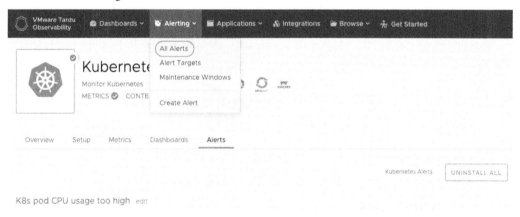

Figure 10.24 – Opening the Alerting page

2. You will see a page as shown in the following screenshot listing the current alert conditions that are already triggered for the configured threshold or under active monitoring:

Figure 10.25 – Alerts page

This concludes our section on how to get started with Aria. We learned how to set up a trial account to get full access to the Aria platform. Then, we integrated an existing Kubernetes cluster with the Aria account with a simple three-step procedure. Then, we saw how to access some useful out-of-the-box dashboards that show insightful details for the cluster with an ability to drill down to get more details. Finally, we saw how to access the default set of alerts that are provided by Aria for Kubernetes integration.

As we know, Aria is an observability tool. The main use case of that kind of tool is to proactively prevent disruptive failures from happening in the first place, and if a failure happens for any reason, finding out the root cause quickly and resolving it. Monitoring dashboards and alerts are the two main capabilities that play key roles for these reasons. Let's learn about them by taking a deeper look, starting with dashboards and charts.

Working with charts and dashboards

In this section, we will cover the following details on using charts and dashboards in Aria:

- Creating new custom charts to visualize any data being sent to Aria
- Creating new custom dashboards using canned or custom charts
- Customizing out-of-the-box dashboards

Let's get started with the topics in this list.

Creating new custom charts

When we configure integration with Aria as we did for a Kubernetes cluster, we can see several metrics getting pushed into Aria that are collected by the agents that we deploy on the hosts. As with a Kubernetes cluster, we can get telemetry data for any integration that we make. The number of metrics and the frequency at which they are ingested in Aria depends on the source and the configurations at the collector and Wavefront Proxy. However, once the metrics data is in the Aria database, it stays there for up to 18 months unless explicitly removed. We can then use that data anytime to build on-demand charts to gain insights into what is going on with a source's health.

The following steps show how to build a chart using an available metric endpoint:

1. Click on the **Dashboard** menu from the top navigation bar and select the **Create Chart** option:

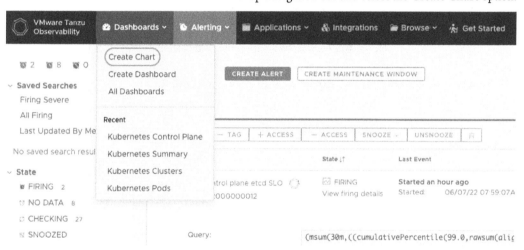

Figure 10.26 – Selecting the Create Chart option

2. On this page, you can directly write a query in either **Wavefront Query Language** (WQL) or **Prometheus Query Language** (PROMQL) structures if you have some knowledge of either of the query languages. If you do not know either of them, Aria allows you to generate a query in the background by selecting the correct values under the dropdowns named **Data**, **Filters**, and **Functions**. Additionally, Aria allows you to either select metrics from the available list or use the integrations you have in place. Since we have a Kubernetes cluster integrated, we can see Kubernetes as an option to build a chart for the Kubernetes integration in the following screenshot. Click on the **Data** dropdown and the **Kubernetes** tile as highlighted in the following screenshot:

Figure 10.27 – Selecting Kubernetes integration

3. Select the chart data points as shown in the following screenshot:

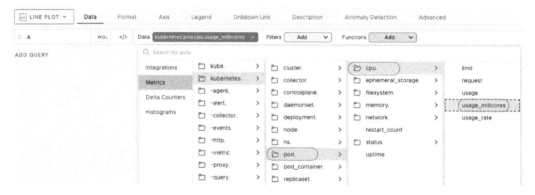

Figure 10.28 – Selecting data elements for the chart

4. The following chart shows a live view of the CPU usage for all Pods belonging to all the namespaces in all the Kubernetes clusters that we have integrated with the Aria account. Hence, this chart would be very clumsy if there are several Kubernetes clusters with several Pods running in them. We may need to implement certain filters to make this chart more usable. In the context of this book, there should only be one Kubernetes cluster, which we integrated earlier in the book:

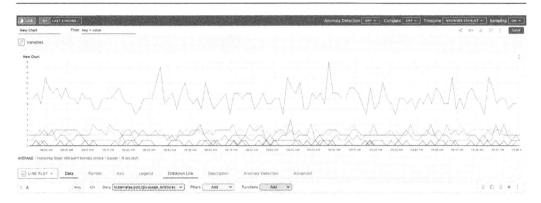

Figure 10.29 – A chart showing the CPU usage of all Pods in all Kubernetes clusters

5. Apply a filter to only see the Pods running in the **wavefront** namespace as shown in the following screenshot. This will show the CPU usage of the Wavefront Collector and Wavefront Proxy Pods that we deployed earlier during the Kubernetes cluster integration steps:

Figure 10.30 – Applying filters for a chart

6. The following screenshot shows a chart only for the Pods under the **wavefront** namespace. We can add more filters by clicking on the + button near the **Filters** dropdown and applying data functions to further fine-tune the chart as per the requirements. Click on the **SAVE** button to save this chart for future use:

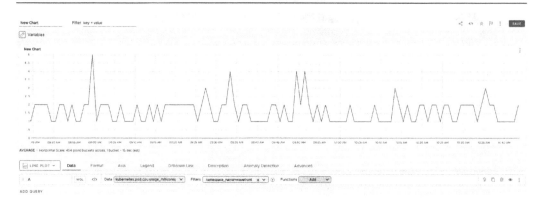

Figure 10.31 – A chart showing CPU usage of all Pods in the wavefront namespace

Now, we have a chart showing some meaningful data on how a group of Pods belonging to a namespace behave in terms of their CPU utilization. We can do the following things with this chart:

- Share it with someone to show them the same view of the data so we can talk in the same language when discussing any performance issues

- Add this chart to either a new or an existing custom dashboard for future use

- Go back in time to see historical performance data for the selected time duration

- Mark interesting events for future reference

- Export the chart in either a PDF or CSV format

- Change the visualization of the chart to get it in other graphical formats such as bar chart, pie chart, tabular chart, and more

- Add more data layers by adding more queries

The following subsection focuses on various user controls that can be applied to a chart in Aria.

Understanding chart manipulation controls

The following screenshot highlights various controls with a corresponding number as a reference:

Figure 10.32 – Chart controls

The controls in the previous screenshot are described against those numbers in the following list:

1. These controls allow you to either watch the chart with live data that keeps refreshing after a few seconds or select a specific timeframe in the past to pull data for that duration only.

2. These controls provide ways to add more details to the chart:

    I.   **Anomaly Detection** finds specific chart patterns that are deemed unusual based on past data using a few artificial intelligence algorithms.

    II. **Compare** lets you visually compare historical data for the same set of sources to find any pattern changes.

    III. **Timezone** lets you change the chart's timeline for the selected time zone. By default, Aria uses the internet browser's time zone.

    IV. **Sampling** lets you turn the usage of data sampling for chart generation in place of considering all the data points on or off.

3. **Filter** allows you to reduce the data points based on the tag values supplied without modifying the chart's query statement.

4. These controls are used for the following reasons, listed from left to right:

    I.   Generate a link to the chart that can be shared with others

    II. Prepare the chart to be embedded in a web page

    III. Show/hide the chart variables if used

    IV. Flag an event point on the chart with details that generates an event log record for future references

    V.   Export the chart as either a PDF or CSV file

    VI. Save the chart in either a new or an existing dashboard

5. These controls allow you to zoom in and out of the chart for the selected time window.

6. This control allows you to change the visualization of the chart to display it in different formats such as pie, bar, table, single-value, and many more.

7. The following list describes these tabs:

    I.   The **Data** tab lets you work with queries to pull chart data

    II.   The **Format** tab allows you to modify the chart's visual details

    III. The **Axis** tab allows you to modify the chart's axis units

    IV.   The **Legend** tab allows you to configure a data legend on the chart

    V.   The **Drilldown Link** tab allows you to configure a link to another dashboard showing relevant data

    VI.   The **Description** tab allows you to set a chart description that can be seen in a tooltip for the chart with mouse-hover

    VII. The **Anomaly Detection** tab allows you to configure related settings for its sensitivity, display, and data sampling

    VIII. The **Advanced** tab allows you to pull obsolete metrics and configure the chart's timeline with respect to its parent dashboard.

8. This control allows you to name a chart data layer for the query.

9. These controls allow you to toggle between WQL and PROMQL, and between the query graphical editor and the text editor.

10. This set of controls allows you to perform the following configurations from left to right:

    I.   See query execution statistics to determine the need for query optimization

    II.   Clone an existing query to create a new data layer for the chart

    III. Delete the query and its respective visual data from the chart

    IV.   Show or hide the query data in the chart without deleting the query

    V.   Various other options, including alert creation based on the query and chart formatting

11. The **ADD QUERY** link allows you to add new data and its visualization layer to the chart.

All these chart controls play a key role in building meaningful dashboards. Let's now learn how to work with dashboards in Aria.

Creating new custom dashboards

As we have seen earlier in this chapter, for Kubernetes integration, Aria supplies a set of canned dashboards that are ready to use for the telemetry data coming from the respective integration source.

However, Aria can ingest data from various integrated sources including applications, Kubernetes, a hypervisor of the virtual infrastructure layer, an OS, databases, and several others. In that case, the default dashboards for a particular integration would not provide a complete picture. For example, none of the default Kubernetes dashboards in Aria cover charts showing data coming from the OS or the hardware hosts on which the nodes are deployed. The default and integration-specific dashboards cannot assume that the data pertaining to another integration will always be present. For example, in this chapter, we have only one integration in place for a Kubernetes cluster. Hence, if a default dashboard expects metrics to come from the node OSs, then it would fail. In these cases, we need custom dashboards that can display correlated data from multiple sources. Dashboards that show data from different sources could significantly help get a broader view of the landscape and quickly find the root causes of issues. A slow-running query in a database could be the cause of a slow-responding application, so having the details of both the layers, the application and the database, on one dashboard could help find the issue quickly.

There are two ways we can build custom dashboards in Aria. First, we can build them using custom charts that can be added to a new or existing dashboard. Second, we can build custom dashboards using the different templates provided in Aria or add custom charts to it. Let's use the chart we built in the previous section to create a new dashboard. The following steps describe this:

1. Click on the **SAVE** button, followed by the **SAVE TO A NEW DASHBOARD** button, as highlighted in the following screenshot:

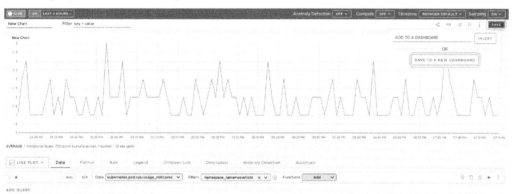

Figure 10.33 – Creating a new dashboard from a custom chart

2. Enter the name of the dashboard and click on the **CREATE** button:

Figure 10.34 – Naming and creating a dashboard

3. The following screenshot shows the Aria dashboard editor where the custom chart is now added. We can add several custom or predefined charts from the out-of-the-box chart templates provided by Aria for the integrations it has:

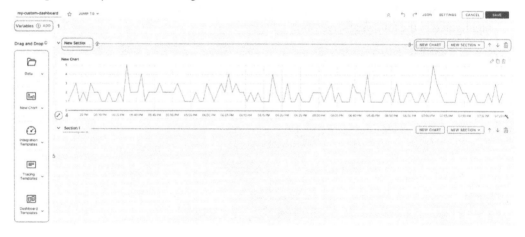

Figure 10.35 – Dashboard editor

The following points explain the highlighted sections in the previous screenshot of the dashboard editor:

1. Variables allowing you to manipulate the data in the applicable charts of the dashboard. These variable values can be added to the chart queries.

2. A section is a group of similar charts in a dashboard. We can use sections to either show/hide charts or jump from one section to another in a very big dashboard.

3. These section controls allow you to add, remove, or move the sections.

4. These arrows allow you to resize the chart to add more than one chart in a single row for more consolidation of data in a row.

5. All these tiles allow you to add more charts to this dashboard in a selected section. Here, we can also import some of the existing canned charts that are provided by Aria for integration.

After seeing how to create a new dashboard from scratch, let's now see how to customize an out-of-the-box dashboard provided by Aria.

Customizing a default dashboard

Although the default dashboards provided in Aria for different integrations are designed thoughtfully, there could be a few things we may like to change in their look and feel. In Aria, we cannot directly update a default dashboard that comes out of the box, but we can clone it and create a custom copy. Then, we can update the custom copy of the default dashboard in whatever way we need. Let's see how that can be done using the following steps:

1. Click on the **All Dashboards** menu under the main **Dashboards** menu in the top navigation bar as shown in the following screenshot:

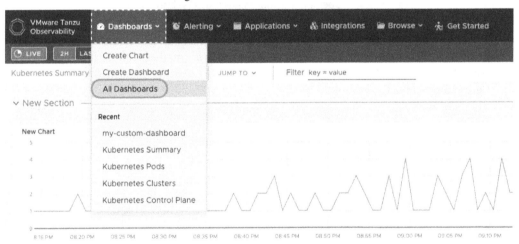

Figure 10.36 – Opening the Dashboards page

2. Click on the **Kubernetes** link from the left-hand navigation bar and then click on the **Kubernetes Clusters** dashboard as highlighted in the following screenshot:

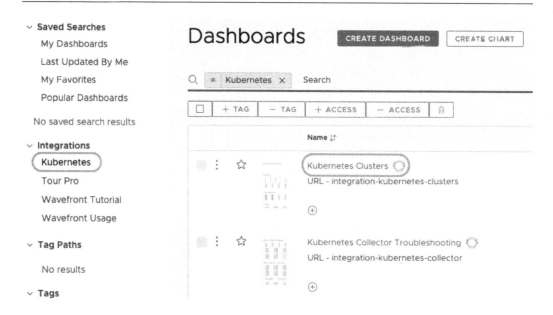

Figure 10.37 – Opening the Kubernetes Clusters dashboard

3. The following screenshot shows the **Kubernetes Clusters** dashboard, which is one of the default dashboards that comes with Aria. All default dashboards are marked with the highlighted lock symbol. To clone the dashboard for customization, click on the highlighted dots menu on the right and select the **Clone** option:

Figure 10.38 – Cloning a default dashboard

4. Give an appropriate name to the cloned copy of the dashboard and click on the **CLONE** button:

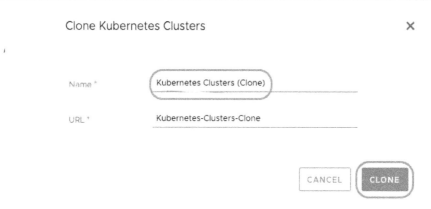

Figure 10.39 – Naming the cloned dashboard

5. You will see the same dashboard editor that we checked out earlier allowing you to make any changes to the dashboard, including but not limited to changing chart data queries, chart visualization, rearrangement of tiles, and adding new data visualizations:

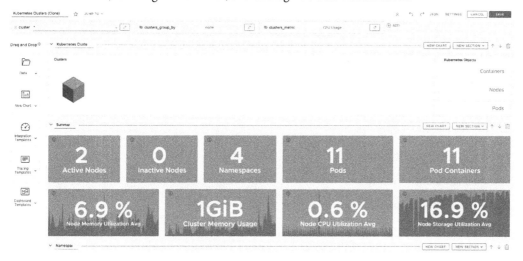

Figure 10.40 – Dashboard editor

Once you make the required changes and save this dashboard, you can then mark it as a favorited dashboard and put it under the quick access menu. You can also define who can access the dashboard for viewing and editing purposes.

With this, we have concluded working with charts and dashboards, with a very high-level overview of the capabilities. For more details about charts and the dashboard, refer to the official product documentation – `https://docs.wavefront.com/ui_charts.html` and `https://docs.wavefront.com/ui_dashboards.html`. In the next and last section of the chapter, we will learn about working with alerts.

Working with alerts

Any monitoring tool is not complete if it does not have a good alerting capability. Aria is also not an exception. It has a very powerful and highly configurable alerting engine that allows you to configure alerts with multiple thresholds, delivery based on those threshold levels, and several possible delivery methods, including but not limited to an email, a text, a Slack message, a Microsoft Teams message, a ServiceNow ticket, or a PagerDuty incident. Moreover, we can also configure triggering an automated process using a webhook that calls a remote API outside of the Aria boundary. There are so many useful things we can configure using Aria's alerting features. Let's learn more about alerts in this section, including the following things:

- Creating alert targets to define who should get which alert using which delivery mechanism
- Defining a maintenance window to avoid obvious alerts
- Creating new alerts
- Inspecting firing alerts

Let's get started with these topics.

Creating alert targets

An alert target in Aria contains several details and provides very flexible ways to define the following things at a high level:

- When the notification should be delivered and when it should be held back
- Using which medium the alert should be delivered
- Who should get the alert and under what conditions
- What details the alert should contain and in what format

The following steps show how to create the most common email-based alert target in Aria:

1. Open the **Alerting** menu from the top navigation bar and select **Alert Targets**:

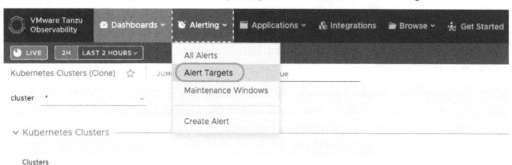

Figure 10.41 – Opening the Alert Targets list page

2.   Click on the **CREATE ALERT TARGET** button:

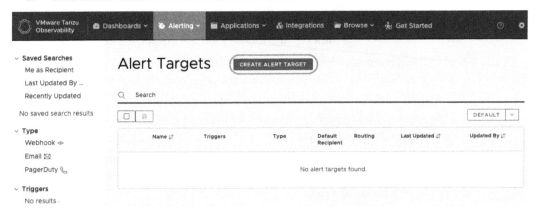

Figure 10.42 – Creating a new alert target

3.   Enter details in the form as depicted in the following screenshot:

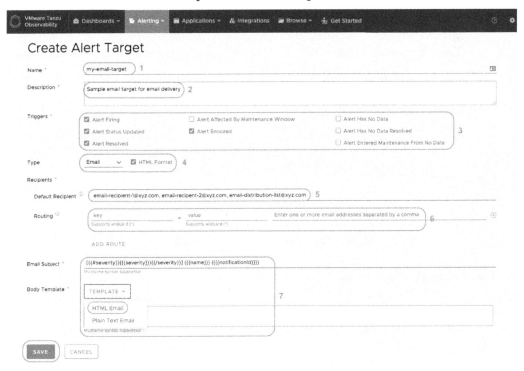

Figure 10.43 – Create Alert Target page

The following are the elements in the previous screenshot as per their numbers:

I.   This is the name of the alert target. Specify a unique name here.

II. This is the description of the alert target. Specify any understandable description.

III. A set of **Triggers** defines when the alert notification should be delivered to its destination and when it should not be. For example, it allows you to configure if you would like to deliver an alert notification when the alert condition is resolved.

IV. The **Type** dropdown allows you to select one of the main three types of alert notification – **Email**, **PagerDuty**, and **Webhook**.

V.   The **Default Recipient** box allows you to specify a list of email addresses that should receive the email notification for this alert target. The email list specified here will get the notification for any dataset.

VI. The **Routing** configuration allows you to deliver the same alert notification to additional destinations based on the defined keys and their values. Using this configuration, we can configure different application teams to be notified based on the application identifiers in the alert data.

VII. These controls allow you to configure the details in the notification. When this pertains to an email notification, the details are the subject and the body of the email. We can configure these values using variables that carry the alert-specific custom data in the notification.

4.   We can see an alert target created as shown in the following screenshot:

Figure 10.44 – Created alert target record

Now, as we have an alert target for the delivery of notifications for any generated alert, we can use this target to deliver alerts. We can use one alert target with more than one alert configuration – but before we learn how to create an alert with a button click, let's first learn how to configure a maintenance window to avoid false-positive alerts during scheduled maintenance when alert monitoring.

Defining a maintenance window

Getting a flood of alerts during a maintenance window for systems under observation is not good, as we already know that things may go down or perform abnormally during those activities. However, in the midst of false positives during a maintenance window, missing a real alert that needs attention would be worse. To prevent critical alerts from being buried under those non-critical alerts, Aria allows you to define a maintenance window where we can mark certain system health parameters to be ignored. In this case, we can take actual alerts seriously during scheduled activities. The following steps show how to configure a maintenance window in Aria:

1. Click on the **Alerting** menu from the top navigation bar and select the **Maintenance Windows** option as shown in the following screenshot:

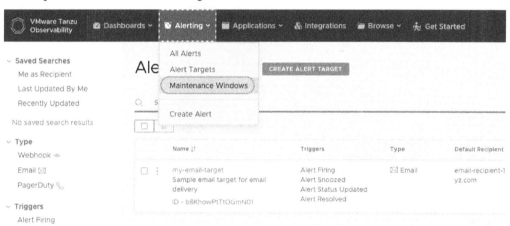

Figure 10.45 – Opening the Maintenance Window list page

2. Click on the **CREATE MAINTENANCE WINDOW** button:

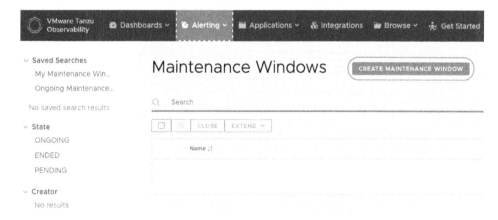

Figure 10.46 – CREATE MAINTENANCE WINDOW

3. Enter the required details of the maintenance window to select the time and the characteristics of the systems for the maintenance window as depicted in the following screenshot:

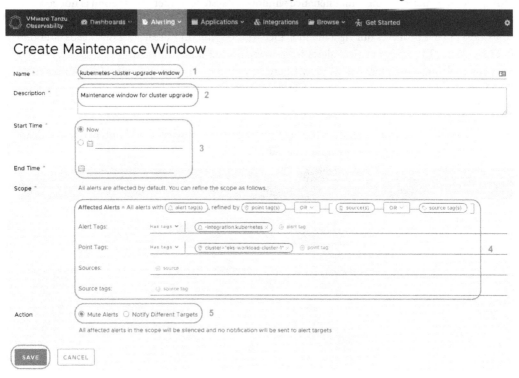

Figure 10.47 – Entering maintenance window details

The following numbered points describe the corresponding number in the previous screenshot:

I.   Enter a unique name for the maintenance window record

II.   Enter a description of the maintenance window's purpose

III.   Either select **Now** to start the maintenance window on demand or select the scheduled start and end time and date to define the window

IV.   Enter the source-defining parameters to cover the alerts in this maintenance window

V.   Select either to mute/ignore the alerts with the criteria covered in the previous point and during the defined time window or select an alert target to divert the alerts there during the window

Once the **SAVE** button is clicked, the maintenance window is created, as shown in the following screenshot:

Figure 10.48 – Created maintenance window record

Now, with the maintenance record created, let's create a custom alert record.

Creating new alerts

In this section, we will learn how to create a new alert. There are three ways we can create alerts in Aria. One of them is to create an alert using the alert creation wizard from the main **Alerts** menu. The second one is to duplicate an alert and modify the required details. The last one is to create it using an existing chart, usually already there on a dashboard. While the first two ways are more obvious, let's learn how to create an alert using the third way, from an existing chart, using the following steps:

1. Open the **All Dashboards** page under the **Dashboards** menu on the top navigation bar:

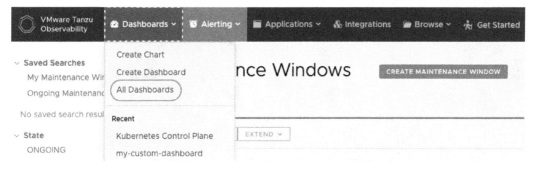

Figure 10.49 – Opening the All Dashboards page

2. Click on the **Kubernetes Clusters** dashboard from the list as shown in the following screenshot:

Figure 10.50 – Opening the Kubernetes Clusters dashboard

3. Click on the pencil icon on the **Node Storage Utilization Avg** chart as highlighted in the following screenshot. The same method of creating an alert from a chart can be applied to all the charts on any dashboard:

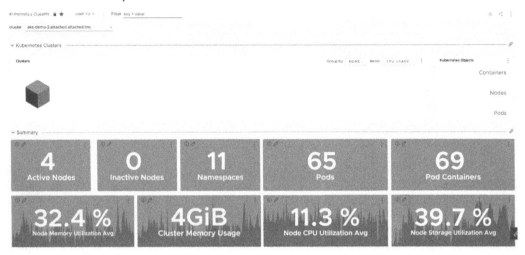

Figure 10.51 – Editing a chart

4. Click on the **Create Alert** link under the dots menu dropdown of the query as shown in the following screenshot:

Figure 10.52 – Creating an alert from a chart editor

As you can see on the **Create Alert** page, the query to pull alert data is already populated using the chart's query. This is a big help in getting the right alert configuration. Let's set up the other alert details.

5. Click the **NEXT** button on the query details page as shown in the following screenshot. Update the alert query if required:

← Create Alert

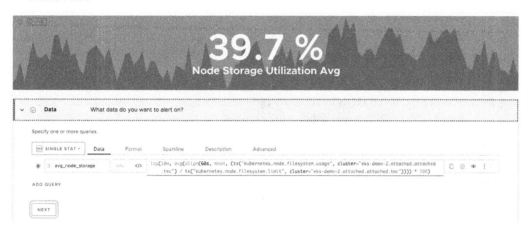

Figure 10.53 – Updating the alert query

6. Configure the alert conditions with different threshold levels as shown in the following screenshot:

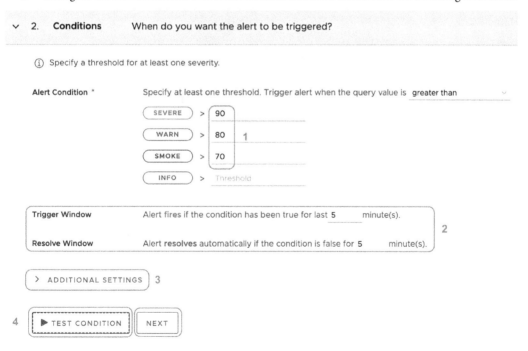

Figure 10.54 – Configuring alert conditions

The following points correspond to the numbers mentioned in the previous screenshot to describe the controls:

I.   These are different alert threshold values. The alert will fire with the corresponding severity when a certain condition is met.

II.   The trigger and resolve windows define the timelines to consider a condition as either the start of the alert or the end of it.

III. The additional settings provide controls to define the frequency of alert checks.

IV.   The **TEST CONDITION** button allows you to test whether an alert would have triggered based on the past data of the last 2 hours to validate the configuration.

7. Select an alert target that has previously been created in the account. Note that the following screenshot only shows one target configured for the **SMOKE** level, which will be applicable for all the threshold levels above it, which are **WARN** and **SEVERE,** too. However, it is also possible to send the alert to different targets for all different threshold levels. Click on the **NEXT** button to move forward:

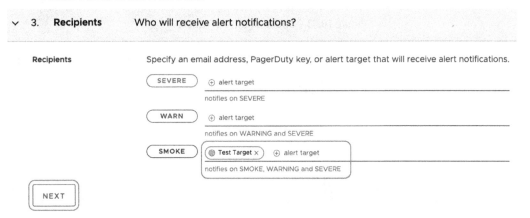

Figure 10.55 – Configuring alert targets

8. Configure alert details that can be helpful for its triage as shown in the following screenshot. Click on the **NEXT** button after that to move forward:

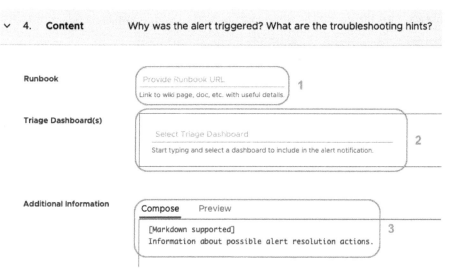

Figure 10.56 – Configuring alert troubleshooting details

The following points describe the controls that are marked with numbers that correspond to the previous screenshot:

I.   In **Runbook**, we can provide an external URL that contains details on how to troubleshoot and resolve the alert condition

II.   Under **Triage Dashboard(s)**, we can add one or more dashboards that will be added to the alert details to provide the required details upfront.

III. Under **Additional Information**, we can provide any arbitrary details that could be helpful for troubleshooting this alert.

9. Give the alert a name and, optionally, the related tag values to identify the sources impacted, and click on the **ACTIVATE** button to create the alert:

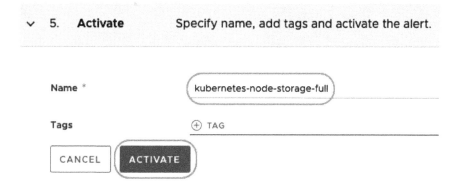

Figure 10.57 – Activating the alert

10. Once activated, the alert can be seen under the **Alerts** page as shown in the following screenshot:

Figure 10.58 – Newly created alert

This concludes the details on creating new alerts in Aria. Learn more about creating alerts in the official product documentation if required: `https://docs.wavefront.com/alerts_manage.html`. Let's now see how to find a firing alert that has an active alert condition.

Inspecting firing alerts

In this section, we will see how to list currently active alerts that are firing with any threshold level configured. The following steps describe this process:

1. Click on the **FIRING** state from the left-hand navigation bar, which will list all the alerts under that status, and click on the name of the firing alert as shown in the following screenshot:

Figure 10.59 – Viewing firing alert details

2. The following screenshot shows a view of the page that contains the details of the firing alerts, including other possible related alert conditions, the chart showing the firing duration, the query used for the calculation, and many other things:

Figure 10.60 – Firing alert details

With this, we conclude the *Working with alerts* section of the chapter. In this section, we started with creating alert targets and learned how we can flexibly define different alert notification types and destinations based on their severity and other attributes of the system under observation. Then, we learned how to create a maintenance window to treat alerts differently during a specific time of scheduled activities that would potentially trigger a bunch of alerts otherwise. After that, we learned how to create an alert using an existing chart from an existing dashboard. The same knowledge can be applied even to creating a new alert from scratch. Lastly, we saw how Aria shows different details of the alerts that are currently firing.

Summary

In this chapter, we learned about a very important component of the Tanzu product portfolio, VMware Aria Operations for Applications by Wavefront. In the first section of the chapter, we gave various reasons explaining the importance of this tool in terms of the observability of modern containerized applications. In this section, we learned about different out-of-the-box integrations supported by Aria. We also discussed how Telegraf collectors and the Wavefront SDK make it possible to integrate almost any data source for observability. Additionally, we discussed other benefits of Aria, including the high-speed data collection and processing power, and the ability to collect different telemetry data types including metrics, events, histograms, and span logs. Then, we listed different categories of data processing functions that can be used with WQL, including data aggregation, filtering, time operations, moving-window aggregation, exponential, and trigonometric.

After that, we learned about various components of the solution. We explored metrics, histograms, span logs, and events. We then discussed the common deployment architecture of the solution and the details of various components with their roles. In that part, we discussed various ways to ingest data in Aria, different types of sources, collectors, Wavefront Proxy, and the Wavefront service, a SaaS platform. Later in the chapter, we learned how to open a trial account for Aria and integrate an existing Kubernetes cluster with the account. Then, we saw how to access a set of default dashboards for Kubernetes integration provided by Aria out of the box. Lastly, we learned about creating custom charts, dashboards, and alerts for the integrations in place.

Although it seems like we covered the solution in great depth, we just scratched the surface. There are still several more capabilities we could not cover in this chapter, as we only covered the details centered around Kubernetes monitoring. Besides the details covered here, Aria has a great set of capabilities for APM using the ingested span logs and application health metrics. We could not even cover topics such as derived metrics, delta counters, user permission configuration, and external logging solution integration because of the limited scope of this chapter.

In the next chapter, we will see how to securely connect various applications deployed in different Kubernetes clusters, even across different cloud infrastructures, using Tanzu Service Mesh.

11
Enabling Secure Inter-Service Communication with Tanzu Service Mesh

Enterprise software has changed significantly in the last few years. Some of you reading this may remember when most software was written as a single monolith by a small team of developers. The application had to be deployed, updated, started, stopped, and scaled as a single unit.

As software began to evolve in the early to mid-2000s, demands on software and software developers started to grow along four axes:

- **Time to value**: Teams were expected to deliver bug fixes, improvements, and new value-delivering features on ever-shrinking timelines.

- **Elasticity**: Teams needed to scale "hot" services independently and deploy individual features without having to build, test, and deploy the entire monolith all at once.

- **Fault-tolerance**: When an individual service failed or started to degrade, it shouldn't affect other services, and the system should be able to work around the problematic service.

- **Programming languages and frameworks**: The technology landscape began to move too quickly for monolithic applications to keep up. Developers felt stifled and constrained by the older, less agile technology.

These demands led to the near-universal adoption of design patterns such as microservices (`https://en.wikipedia.org/wiki/Microservices`). Furthermore, as microservices came to predominate software development and teams became more independent and autonomous, new architecture paradigms began to emerge, specifically hybrid and multi-cloud:

- **Hybrid cloud**: This refers to deploying apps consisting of legacy VMs, Kubernetes services, serverless functions, and possibly other technologies such as **Platform-as-a-Service** (**PaaS**) across an on-premises data center and the public cloud.

- **Multi-cloud**: This refers to deploying an application across multiple **Virtual Private Clouds** (**VPCs**) across one or more cloud providers.

Google has some interesting material on these architectures here: `https://cloud.google.com/architecture/hybrid-and-multi-cloud-patterns-and-practices`.

For reasons we'll discuss shortly, Tanzu Service Mesh is the ideal tool for teams tasked with securely and consistently delivering fixes and features while operating in complex modern environments.

In this chapter, we will cover these topics:

- Why Tanzu Service Mesh?
- Features and capabilities of Tanzu Service Mesh
- How to get started with Tanzu Service Mesh
- How to perform key day-2 operations on Tanzu Service Mesh
- GSLB with NSX-T Advanced Load Balancer and Tanzu Service Mesh

With that, let's jump in and dive deep into why a team might need Tanzu Service Mesh.

Why Tanzu Service Mesh?

Tanzu Service Mesh is a tool built expressly to enable the meaningful business outcomes we just described (fast time to value, elasticity, fault tolerance, and so on) while operating in the context of a hybrid or multi-cloud environment. Here are some examples:

- How do we deliver a faster time to value when we have to update 100 dependent services whenever a core service moves from one cloud to another?

- Do we realize the full value of an elastic system if every cloud has a different process and toolset for deploying and scaling?

- The same goes for detecting and remediating faults. Are we fully benefitting from a resilient system if we have to maintain different tools for monitoring, alerting, and remediating across clouds?

- Let's say that you were to invest the time and effort to centralize tooling and monitoring to work across multiple clouds. How do you keep your tech up to date as languages, frameworks, and technologies continue their inexorable forward march? Is the continued ongoing investment worthwhile?

Tanzu Service Mesh sits squarely in the middle of this problem space, bringing the benefits of large, distributed microservice architectures to multi and hybrid cloud architectures. It may help to visualize the problem space: delivering business outcomes across clouds and on-premises, across multiple technologies and frameworks:

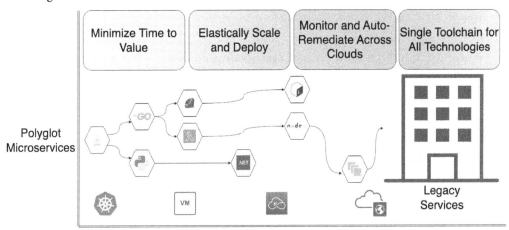

Figure 11.1 – Business outcomes across clouds and technologies

Now that we've scratched the surface of why you would want to use this tool, let's dive deep into the features and capabilities that enable those *whys*.

Features and capabilities of Tanzu Service Mesh

Here are eight concrete features that Tanzu Service Mesh provides. These features enable the specific capabilities necessary to accomplish what we described in the previous section: how we deliver tangible outcomes in a multi or hybrid cloud environment:

- **Out-of-the-box enablement of hybrid and multi-cloud architectures**: With a few clicks in the UI or API calls, you can onboard multiple different clouds and architectures (Kubernetes, legacy, serverless) into a single virtual space where you can quickly and easily stand up any of the various industry standard multi-cloud or hybrid cloud architectures.

- **Effortlessly move apps between clouds**: You can deploy your app in a different VPC, a different platform (VMs versus Kubernetes), or even an entirely different cloud provider without affecting dependent apps and services.

- **Automatic high availability**: You can deploy the same service to multiple clouds and the service mesh will automatically load balance between them. This includes transparently failing over from one cloud to another when the service fails in one location.

- **SLO tracking with AutoScaling**: Tanzu Service Mesh allows you to define **Service-Level Objectives (SLOs)**, track your error budget relative to the SLOs, and even autoscale services when SLOs aren't being met.

- **End-to-end mTLS**: With zero operator effort or toil, Tanzu Service Mesh uses sidecars to secure all inter-service traffic in both directions. This allows for guaranteed client and server identity, which can be used to further secure the services.

- **Security policies and auditing**: With bi-directional TLS, client and server IDs are cryptographically guaranteed. This allows you to create airtight *allow* and *deny* policies that dictate which services can communicate with each other.

- **Full workload visualization**: Tanzu Service Mesh gives a real-time visualization of all the services in the mesh, complete with metrics.

- **Tight control over deployments**: Tanzu Service Mesh gives you all the tools in the Istio toolbox to control rollouts of a service: traffic shaping, canary deploys, A/B testing, and more.

With those features, I hope you now understand why you would use Tanzu Service Mesh. Now, let's get our hands dirty and get started with it.

How to get started with Tanzu Service Mesh

Tanzu Service Mesh consists of controllers that install into your Kubernetes clusters as well as a central SaaS global control plane. The SaaS control plane does a few things:

- Manages installation of Tanzu Service Mesh components onto Kubernetes clusters

- Automatically updates those components as necessary

- Sends global configuration down to those controllers (for example, the location of services on other clusters)

- Manages and deploys policies on the Kubernetes clusters

- Gathers metrics to enable visualization and SLOs

To get started with Tanzu Service Mesh, let's log into our VMware Cloud Services Console and select the **Tanzu Service Mesh** tile from the list of services.

> **Important note**
>
> Tanzu Service Mesh is the one product in this book that's neither free to use nor provides a self-service avenue to enable a trial. If you don't currently have a license for Tanzu Service Mesh, you'll need to reach out to your VMware Account Executive to set up a trial. If that's not possible, much of what we'll cover in this section is also included in a free VMware hands-on lab for Tanzu Service Mesh located here: `https://labs.hol.vmware.com/HOL/catalogs/lab/8509`.

The VMware Cloud Services Console is located here: `https://console.cloud.vmware.com/csp/gateway/portal/#/consumer/services/organization`.

Select **Launch Service** from the **Tanzu Service Mesh** tile, as shown in the following screenshot:

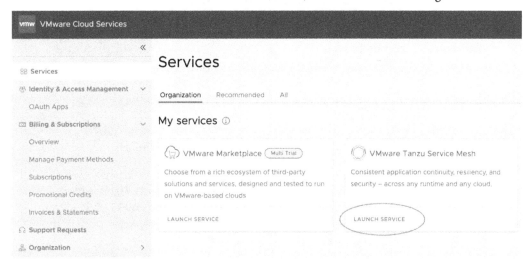

Figure 11.2 – Launching Tanzu Service Mesh

Once we launch the Tanzu Service Mesh app, we're ready to onboard our Kubernetes clusters.

Onboarding Kubernetes clusters

This example requires two separate clusters with the ability to create `LoadBalancer` services. You can refer to the *Appendix* to explore your options for standing up Kubernetes clusters. For this exercise, I'd strongly recommend using one of the public cloud offerings (EKS, AKS, GKE, or TKG on the public cloud) simply because they make it very easy to stand up `LoadBalancer` services, which are a must-have for this exercise.

Here are the steps to onboard your clusters:

1. Select **New Workflow** | **Onboard New Cluster**.

2. In the dialog, give your cluster a name (for example, `cluster-1`) and click **Generate Security Token**. This will generate some credentials that your Kubernetes cluster will use to connect to the SaaS control plane.

3. From your terminal, make sure that you have kubectl pointed to your first cluster:

```
ubuntu@ip-172-31-33-59 ~> kubectl config get-contexts
CURRENT   NAME          CLUSTER              AUTH-
INFO                    NAMESPACE
*             user@eks-tsm-1.us-west-2.eksctl.
io                     eks-tsm-1.us-west-2.eksctl.
io                               acmedemo-tkg@eks-tsm-1.
us-west-2.eksctl.io
              user@eks-tsm-2.us-west-2.eksctl.
io                     eks-tsm-2.us-west-2.eksctl.
io                               acmedemo-tkg@eks-tsm-2.
us-west-2.eksctl.io
```

4. Then, copy each of the kubectl commands from the UI into your terminal:

```
ubuntu@ip-172-31-33-59 ~> kubectl apply -f  'https://
prod-4.nsxservicemesh.vmware.com/tsm/v1alpha2/projects/
default/clusters/onboarding-manifest?tenant=b011ef56-
1670-4d99-9179-0000000000000'
namespace/vmware-system-tsm created
customresourcedefinition.apiextensions.k8s.io/
aspclusters.allspark.vmware.com created
customresourcedefinition.apiextensions.k8s.io/clusters.
client.cluster.tsm.tanzu.vmware.com created
customresourcedefinition.apiextensions.k8s.io/
tsmclusters.tsm.vmware.com created
customresourcedefinition.apiextensions.k8s.io/
clusterhealths.client.cluster.tsm.tanzu.vmware.com
created
configmap/tsm-agent-operator created
serviceaccount/tsm-agent-operator-deployer created
```

```
clusterrole.rbac.authorization.k8s.io/tsm-agent-operator-
cluster-role created
role.rbac.authorization.k8s.io/vmware-system-tsm-
namespace-admin-role created
clusterrolebinding.rbac.authorization.k8s.io/tsm-agent-
operator-crb created
rolebinding.rbac.authorization.k8s.io/tsm-agent-
operator-rb created
deployment.apps/tsm-agent-operator created
job.batch/update-scc-job created


ubuntu@ip-172-31-33-59 ~> kubectl -n vmware-system-
tsm create secret generic cluster-token --from-
literal=token=eyJhb…
secret/cluster-token created
```

The first command installs the control plane components. These components will turn around and install Istio locally, as well as check in with the global SaaS control plane to check for updates.

The second command creates a Kubernetes secret that will be used to authenticate to the global SaaS control plane.

- Finally, click the green **Install Tanzu Service Mesh** button. That will instruct the components running on your Kubernetes cluster to pull down and install the Istio data plane components.

- Repeat this entire process for your second Kubernetes cluster.

At this point, you should have Tanzu Service Mesh up and running on both of your clusters and they should be visible in the Tanzu Service Mesh UI in the **Clusters** pane. You should see something like what's shown in the following screenshot. You'll notice that we have four nodes on each cluster but no services yet:

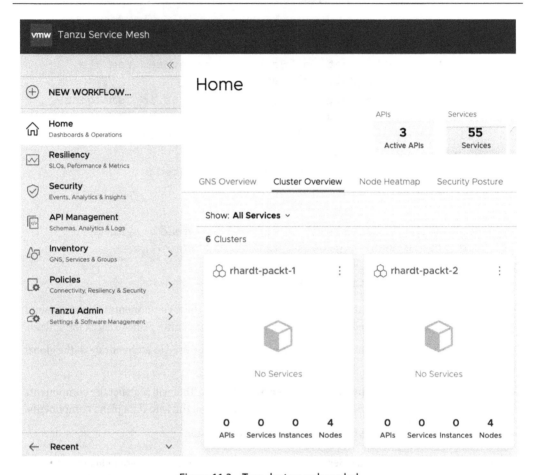

Figure 11.3 – Two clusters onboarded

Next, we'll need to create a namespace in each cluster that we'll use as our Global Namespace in Tanzu Service Mesh.

Creating a Tanzu Service Mesh Global Namespace

A **Global Namespace** (**GNS**) is a logical namespace in Tanzu Service Mesh that can span multiple clusters. In our case, we'll create a Global Namespace called acme in each cluster. Now, all the services in the acme namespace on cluster 1 can see all the services in the acme namespace on cluster 2, and vice versa. Here's how we go about creating the GNS across our two onboarded clusters:

1. Go ahead and create the namespaces:

    ```
    ubuntu@ip-172-31-33-59 ~> kubectl config
    use-context   acmedemo-tkg@eks-tsm-1.us-west-2.eksctl.io
    ```

```
# target the 1st cluster
Switched to context "acmedemo-tkg@eks-tsm-1.us-west-2.
eksctl.io".
ubuntu@ip-172-31-33-59 ~> kubectl create ns acme
namespace/acme created
ubuntu@ip-172-31-33-59 ~> kubectl config
use-context  acmedemo-tkg@eks-tsm-2.us-west-2.eksctl.io #
target the 1st cluster

Switched to context "acmedemo-tkg@eks-tsm-2.us-west-2.
eksctl.io".
ubuntu@ip-172-31-33-59 ~> kubectl create ns acme
namespace/acme created
```

2. Next, we must create the logical GNS in the Tanzu Service Mesh UI. Here are the steps at a high level:

I.   Select **New Workflow | New Global Namespace**.

II.   Enter the name of the Global Namespace. I used `acme-gns`.

III. Select a domain for the GNS. This will be how services find each other, so I'd recommend against something that may need to resolve via DNS (for example, `acme.com`, `mygns.net`). Instead, I chose to use `devsecops-acme.gns` as something that won't be confused with a public domain name. For future steps to work without additional changes, I'd recommend using the same domain for your GNS.

IV.   You should see something like the following:

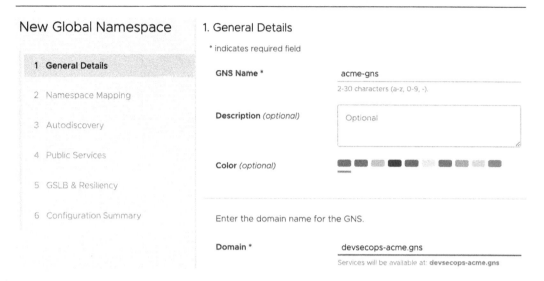

Figure 11.4 – Naming and selecting a unique domain

V.   Next, we'll need to map the acme namespace within each of our two clusters to this GNS on the **Namespace Mapping** screen, as per the following screenshot:

Figure 11.5 – Global Namespace Mapping

VI.   For the remaining screens, you can accept the defaults. In the end, you should see your GNS in the **GNS Overview** tab in the UI:

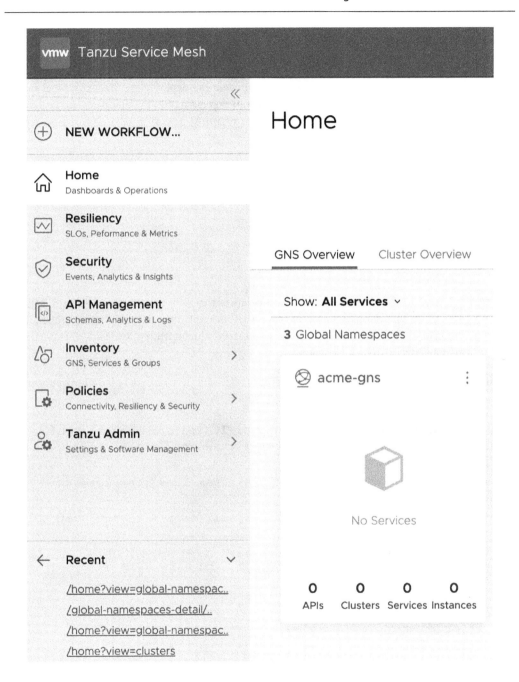

Figure 11.6 – New GNS

Now that our GNS is up and running, it's time to deploy some services to our clusters that will communicate over the GNS.

Installing services

The files in the following steps will install a simple microservice-based application called *Acme Fitness*. If you chose devsecops-acme.gns as your domain, you can deploy these files as-is; otherwise, you'll need to update the internal references to reflect your domain name. The manifests can be found here: https://github.com/PacktPublishing/DevSecOps-in-Practice-with-VMware-Tanzu/tree/main/chapter-12:

Make sure you're pointing to your first Kubernetes cluster and deploying the YAML manifest for cluster-1. It's important to note -n acme at the end of the commands. You need to explicitly declare that these services are to be deployed to the acme namespace. This will install a few things:

- Secrets that apps will use to talk to their data services.

- Several data services (MongoDB, Redis, and so on).

- Some microservice-based applications.

- A gateway, which is like a virtual L7 load balancer that tells Tanzu Service Mesh where to route certain incoming requests. In our case, this is the only external-facing application running on the cluster, so we'll route all requests coming in from outside the cluster to our frontend service, which we have called shopping.

Then, switch to cluster-2 and deploy that manifest. Here are the commands I used to deploy the services into both of my Kubernetes clusters:

```
ubuntu@ip-172-31-33-59 ~/c/D/chapter-12 (main)> kubectl config
use-context  acmedemo-tkg@eks-tsm-1.us-west-2.eksctl.io #
target the 1st cluster
Switched to context "acmedemo-tkg@eks-tsm-1.us-west-2.eksctl.
io".
ubuntu@ip-172-31-33-59 ~/c/D/chapter-12 (main)> kubectl get all
-n acme
No resources found in acme namespace.
ubuntu@ip-172-31-33-59 ~/c/D/chapter-12 (main)> kubectl apply
-f ./cluster1.yaml -n acme
secret/redis-pass created
secret/order-mongo-pass created
secret/users-mongo-pass created
gateway.networking.istio.io/acme-gateway created
virtualservice.networking.istio.io/acme created
service/cart-redis created
deployment.apps/cart-redis created
```

```
service/cart created
deployment.apps/cart created
service/shopping created
deployment.apps/shopping created
service/order-mongo created
deployment.apps/order-mongo created
service/order created
deployment.apps/order created
service/payment created
deployment.apps/payment created
configmap/users-initdb-config created
service/users-mongo created
deployment.apps/users-mongo created
service/users created
deployment.apps/users created
ubuntu@ip-172-31-33-59 ~/c/D/chapter-12 (main)> kubectl config
use-context  acmedemo-tkg@eks-tsm-2.us-west-2.eksctl.io #
target the 2nd cluster
Switched to context "acmedemo-tkg@eks-tsm-2.us-west-2.eksctl.
io".
ubuntu@ip-172-31-33-59 ~/c/D/chapter-12 (main)> kubectl apply
-f ./cluster2.yaml -n acme
secret/catalog-mongo-pass created
configmap/catalog-initdb-config created
service/catalog-mongo created
deployment.apps/catalog-mongo created
service/catalog created
deployment.apps/catalog created
ubuntu@ip-172-31-33-59 ~/c/D/chapter-12 (main)>
```

Accessing the application

Now that we've installed the services across two clusters, let's verify that the app is up and running. Switch your kubectl context back to the first cluster and get the details of the istio-ingressgateway service in the istio-system namespace. If you're running on EKS, you might get the hostname of a load balancer; otherwise, you may see an IPv4 address. Grab the hostname or external IP and plug it into your browser using http, not https.

Here's how I got the service. Notice that I'm pointed at the first cluster and I'm on EKS, so I'm getting a load balancer hostname as my *external IP*:

```
ubuntu@ip-172-31-33-59 ~/c/D/chapter-12 (main)> kubectl get svc
-n istio-system istio-ingressgateway
NAME                      TYPE            CLUSTER-IP      EXTER-
NAL-IP                    PORT(S)                         AGE
istio-ingressgateway      LoadBal-
ancer     10.100.7.70     a075e7d12da464c16a17e0aa0435fd-
be-2ff7515fc9cc6551.elb.us-west-2.amazonaws.com     15021:30588/
TCP,80:31893/TCP,443:31157/TCP     22h
ubuntu@ip-172-31-33-59 ~/c/D/chapter-12 (main)>
```

And here's what you'd expect to see in the browser. If the yoga mat and water bottle images appear, then you know everything is working. The page itself is hosted in cluster 1 and the catalog is hosted in cluster 2!

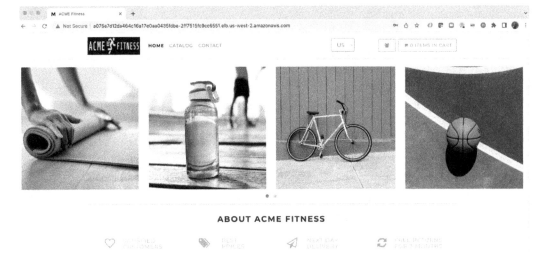

Figure 11.7 – Acme Fitness deployed across two clusters

Next, we'll need to generate some inter-service traffic so that the Tanzu Service Mesh UI can visualize all the connections.

Generating application traffic

Here are some things you can do to make sure that all the services get some traffic:

- Click the login button next to the shopping cart.
- Log in with the following credentials – username: `dwight` password: `vmware1!`.

- Click the **Catalog** link at the top of the screen.

- Browse to an item.

- Add it to your cart.

- Click the **Shopping Cart** button.

- Click **Proceed to Checkout**.

- Go through the checkout process. Any 16-digit number will work for a credit card number.

Now, we can return to the GNS screen for our `acme-gns` in the Tanzu Service Mesh UI. You should see all the services spread across the two clusters, as shown in the following screenshot:

Figure 11.8 – Tanzu Service Mesh service visualization

At this point, you have successfully done the following:

- Stood up a new Service Mesh

- Onboarded multiple clusters

- Federated them into a GNS

- Deployed a distributed app

- Verified the app's functionality and cross-cluster communications

That's an impressive list for day one. Now, let's move on to some operations you may consider after you've accomplished your day-1 checklist.

How to perform key day-2 operations on Tanzu Service Mesh

At this point, we have a distributed application successfully working across two Kubernetes clusters. If you were doing this in a real production environment, there are some *day-2* concerns you'd want to address. For example, it's great that the catalog service can run on a separate cluster, but what if that cluster goes down? How could we load balance across instances on multiple clusters?

Furthermore, we're living in a world where deployment, upkeep, measuring, and monitoring of services are often the responsibility of **Site Reliability Engineers** (**SREs**). If you were the SRE for Acme Fitness, you would have already identified **Service-Level Indicators** (**SLIs**) and defined SLOs for your services. Tanzu Service Mesh greatly simplifies this by allowing you to define your SLIs and SLOs right in the Tanzu Service Mesh UI and measure how your services are meeting those SLOs using real-time metrics.

Enabling service high availability

First, let's address making a service highly available. In our example app, most of the services are on `cluster-1`, while the catalog and its data store are on `cluster-2`. Tanzu Service Mesh lets us stand up additional instances of a service across multiple clusters and automatically load balances between all the instances as they come online.

For our purposes, let's just deploy another instance of the catalog service onto `cluster-1`. We can do that very easily by targeting `cluster-1` and applying the `cluster-2` deployment manifest, like so:

```
ubuntu@ip-172-31-33-59 ~/c/D/chapter-12 (main)> kubectl config
use-context  acmedemo-tkg@eks-tsm-1.us-west-2.eksctl.io #
target cluster 1
Switched to context "acmedemo-tkg@eks-tsm-1.us-west-2.eksctl.
io".
ubuntu@ip-172-31-33-59 ~/c/D/chapter-12 (main)> kubectl apply
-f ./cluster2.yaml -n acme # this will land the catalog service
on cluster 1
secret/catalog-mongo-pass created
configmap/catalog-initdb-config created
service/catalog-mongo created
deployment.apps/catalog-mongo created
service/catalog created
deployment.apps/catalog created
ubuntu@ip-172-31-33-59 ~/c/D/chapter-12 (main)>
```

Now, if we go back to the app and access the catalog a few times, we should see a visualization like the one shown in the following screenshot. Notice how Tanzu Service Mesh automatically load balances between the two catalog instances: one in `cluster-1` and the other in `cluster-2`:

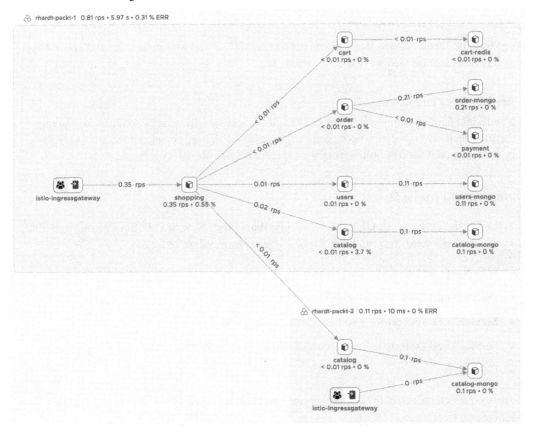

Figure 11.9 – Automatic load balancing across multiple clusters

Next, let's track some SLOs.

Defining and measuring SLOs

For this task, we'll walk through the desired outcome and the steps we follow to achieve that outcome. We'll show some screenshots along the way so we can double-check our work. Then, we'll talk about what the various fields represent.

Desired outcome

Let's say that we're the SREs for Acme Fitness and we have an SLO for the catalog service that it be up and healthy 99.99% of the time, sometimes called *four nines uptime*. In our case, we define *healthy* as having 90% of our requests complete within 100 milliseconds.

Let's go into the Tanzu Service Mesh UI and define this SLO so that we can have Tanzu Service Mesh accurately measure how well we're meeting this SLO.

> **Site Reliability Engineering**
>
> If you're not familiar with the concepts of SREs, SLOs, and SLIs, I'd strongly recommend reading the *Site Reliability Engineering* book. You can get an eBook or hard copy from the usual sources, or you can read it for free online from the Google SRE website: `https://sre.google/`.

Implementing the SLO

From the Tanzu Service Mesh UI, navigate to **New Workflow** | **New SLO Policy** | **Monitored SLO**. This will launch a wizard. You can fill in the first screen's form values as follows:

- **SLO Name**: `acme-catalog-health`
- **Policy Scope**: `GNS Policy, GNS=acme-gns`
- **Service Level Indicators**: p90 Latency is less than 100 ms
- **Service Level Objective**: The service should be healthy 99.99% of the time
- Click **Next**

On the second screen, select the *catalog* service and click next. You could choose to apply this SLO to multiple services if you want to.

Finally, you'll be shown a summary like this one. Notice that it calculates the approximate monthly budget of time your service can be outside of the healthy range. As the service goes into an unhealthy state over a month, the budget will be updated:

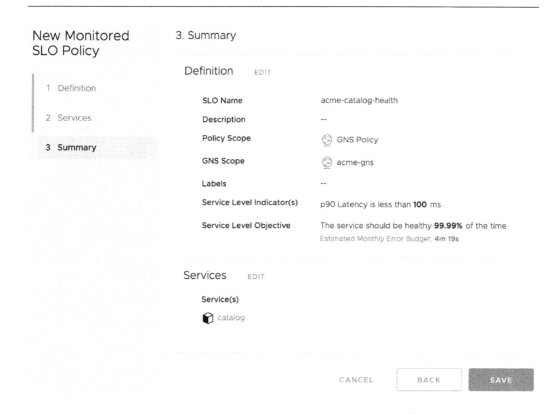

New Monitored SLO Policy

1 Definition

2 Services

3 Summary

3. Summary

Definition   EDIT

SLO Name	acme-catalog-health
Description	--
Policy Scope	GNS Policy
GNS Scope	acme-gns
Labels	--
Service Level Indicator(s)	p90 Latency is less than **100** ms
Service Level Objective	The service should be healthy **99.99%** of the time Estimated Monthly Error Budget: **4m 19s**

Services   EDIT

Service(s)

catalog

CANCEL   BACK   SAVE

Figure 11.10 – SLO details

Now, you can view your services' performance relative to their SLOs by navigating to **Policies | SLOs** and clicking on the name of your SLO. Here's one I set up previously with an unreasonable latency threshold of 5 ms. This was to demonstrate what things look like when you don't meet your error budget:

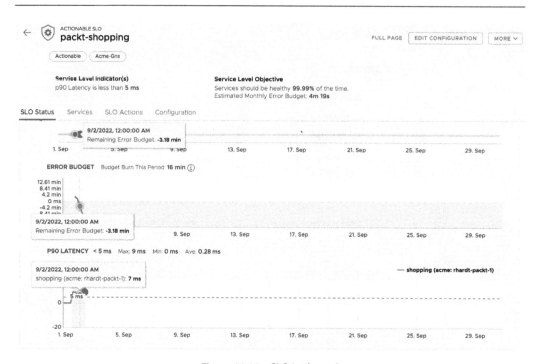

Figure 11.11 – SLO in the red

This shows us that we have an error budget of 4 minutes and 19 seconds for the month to be in an unhealthy state, and we've already exceeded that budget by just over 3 minutes. SLOs are incredibly powerful tools for SREs and they're very easy to set up and maintain in Tanzu Service Mesh. Now, let's move on to some other day-2 tasks that might be of interest.

Other day-2 operations for further research

To comprehensively cover all the day-2 functionality wrapped up in Tanzu Service Mesh would potentially double the length of this book; so, with that in mind, here's a list of some items for further exploration, along with links to the relevant documentation:

- **Actionable SLOs**: You can configure your SLO to perform auto-healing actions when an SLI falls out of range. For example, when latency gets too high, Tanzu Service Mesh can autoscale the service to better handle the load: `https://docs.vmware.com/en/VMware-Tanzu-Service-Mesh/services/slos-with-tsm/GUID-1B9A2D61-D264-44FB-8A06-40277AD42A8E.html`.

- **Onboard some External Services**: Some services live outside of Kubernetes but need to participate in the mesh: `https://docs.vmware.com/en/VMware-Tanzu-Service-Mesh/services/using-tanzu-service-mesh-guide/GUID-F7DC3814-0C3B-42E8-94A1-64B4B182D783.html`.

- **Create a Public Service**: This is a service that is reachable from the internet or a network outside the scope of the service mesh, such as an on-premises corporate network: `https://docs.vmware.com/en/VMware-Tanzu-Service-Mesh/services/using-tanzu-service-mesh-guide/GUID-58A3FA7C-4EFC-44B2-B37B-D2152CB16359.html`.

Now that we've thoroughly covered some day-2 tasks, let's dive deep into one particularly important one, **Global Server Load Balancing (GSLB)**.

GSLB with NSX-T Advanced Load Balancer and Tanzu Service Mesh

One particular day-2 task that is worthy of its own section is implementing GSLB with Tanzu Service Mesh and NSX Advanced Load Balancer.

NSX Advanced Load Balancer

NSX Advanced Load Balancer (NSX-ALB) is a product that, while not technically part of the Tanzu portfolio, complements it very well. You can learn more about it, especially as it pertains to GSLB, here: `https://avinetworks.com/docs/20.1/gslb-architecture-terminology-object-model/#gslbsites`.

In short, NSX-ALB is a powerful software load balancer that's completely API-driven, allowing for tight integrations with other API-driven technologies such as Kubernetes or Tanzu Service Mesh. In addition to enabling the usual functions of an enterprise load balancer such as VIPs, traffic policies, and WAF, it also provides a robust DNS implementation, which makes it especially useful for GSLB. The following is a rough diagram of how Tanzu Service Mesh and NSX-ALB work alongside your corporate DNS (or public DNS) to provide access to services across data centers, multiple regions, and different cloud providers:

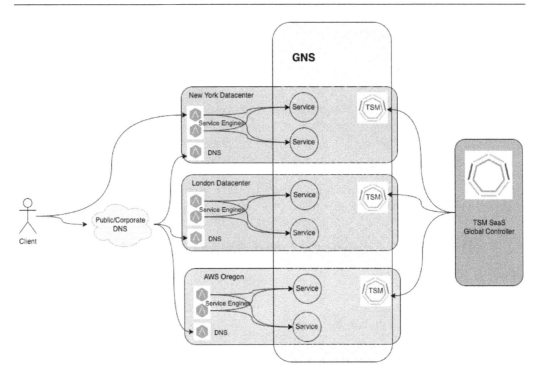

Figure 11.12 – NSX Advanced Load Balancer and Tanzu Service Mesh

Before we can fully understand how Tanzu Service Mesh works with NSX-ALB's GSLB, let's quickly describe how NSX-ALB does GSLB on its own.

Detour – GSLB without Tanzu Service Mesh

NSX ALB enables GSLB by implementing a DNS that will return the IP of a working instance of the service. Here's the sequence of events:

1. The client asks their local DNS server for an IP address for shopping.gslb.acme.com.

2. The local DNS forwards to the authoritative DNS server for gslb.acme.com, which lives in the New York data center.

3. The authoritative DNS server, depending on the GSLB load balancing strategy and the health of the service instances, will return one of three IP addresses: the instance in the New York data center, the instance in the London data center, or the instance in AWS Oregon. Let's say it returns 20.30.40.50, which is the IP address of the NSX ALB Service Engines in the New York data center.

4. The client makes an HTTPS request to 20.30.40.50 with an SNI header of shopping. gslb.acme.com.

5. The NSX ALB Service Engine knows that requests with that particular SNI get routed to the `shopping` service in the `acme` namespace in the Kubernetes cluster running locally in the data center.

6. The client successfully orders his yoga mat.

Now, let's layer in Tanzu Service Mesh.

GSLB with NSX-ALB and Tanzu Service Mesh

Everything from the previous section still applies, but rather than an operator manually setting up GSLB for the `shopping` service, she registers it in the service mesh as a `public` service (`https://docs.vmware.com/en/VMware-Tanzu-Service-Mesh/services/using-tanzu-service-mesh-guide/GUID-58A3FA7C-4EFC-44B2-B37B-D2152CB16359.html`).

Then, she integrates Tanzu Service Mesh with her NSX ALB instances. Finally, Tanzu Service Mesh automatically begins to configure GSLB. Wherever the `shopping` service is deployed in the GNS, Tanzu Service Mesh will configure the GSLB DNS resolver to point to a cluster containing that service. You can read about this process in detail here: `https://docs.vmware.com/en/VMware-Tanzu-Service-Mesh/services/using-tanzu-service-mesh-guide/GUID-896EA3FA-17EA-49E2-B981-5C9634E45512.html`.

Summary

In this chapter, we onboarded onto Tanzu Service Mesh, deployed a real-world application, discussed day-2 operations, and even covered GSLB, a very advanced topic. In the next chapter, we'll attempt to put all the pieces together and paint a picture of what "good" looks like when using the entire Tanzu portfolio. I hope you'll continue!

12
Bringing It All Together

Congratulations on coming this far to the last chapter of the book! In all the previous chapters, we learned how different products inside the VMware Tanzu portfolio solve the challenges of modern application development, security, and operations with the outlined DevSecOps capabilities of Tanzu. This book was written to provide its readers with an introduction to the overall Tanzu capabilities so that they could appreciate what they could do with the whole set of products. VMware has done a great job with its vision for Tanzu to provide a one-stop shop for everything you need for your modern apps. As you can rightly imagine, there is a lot more in this portfolio that we did not cover in this book. Additionally, we just brushed the surface with regard to many of the products covered in this book. The idea was to provide enough details to understand the position of each product and help the reader play with them for learning purposes. This book could be the first stepping stone to begin a more in-depth learning journey for the products that are applicable to your real-life use cases.

As per the outline, we will cover a broader picture of Tanzu and its adoption options. We will also attempt to foresee the future of this ecosystem, so, let's get started to get the book finished.

In this chapter, to bring it all together, we're going to cover the following main topics:

- Tanzu adoption options
- Tanzu beyond this book
- The end-to-end picture
- Pros and cons of a single-vendor solution
- Future of Kubernetes
- What is next for Tanzu?

Tanzu adoption options

At the time of writing, most of the products discussed in this book are standalone and can be consumed with à la carte pricing. However, VMware has bundled them together to provide a simplified and more cost-effective adoption experience. The following is the list of the bundles presently available that enterprises may choose from.

Tanzu Standard

This is the smallest commercially available Tanzu bundle to provide the minimal tools to deploy a multi-cloud Kubernetes platform. It mainly includes Tanzu Kubernetes Grid with essential capabilities of Tanzu Mission Control. Tanzu Standard also includes another flavor of Kubernetes offering that is specific to vSphere. Here, Kubernetes is integrated inside vSphere to make its adoption simpler for the virtual infrastructure admins to get a good understanding of the vSphere ecosystem.

Tanzu Application Platform (TAP)

- We covered this product bundle in depth in *Chapter 8, Enhancing Developer Productivity with Tanzu Application Platform*. This bundle includes products that simplify the life of developers for building and deploying applications on a Kubernetes platform. As of writing this book, this product bundle is supported on Tanzu Kubernetes Grid, Amazon's **Elastic Kubernetes Service (EKS)**, **Azure Kubernetes Service (AKS**, and **Google Kubernetes Engine (GKE)** as a multi-cloud solution. While TAP includes several open source and individually available Tanzu products, the following list contains some of the major ones in this bundle:

- **Tanzu Application Accelerator** – to define ready-to-use application templates with good defaults and required boilerplate code

- **Tanzu Build Service** – to build application container images from the source code using the Cloud Native Buildpacks open source project

- **Application Live View** – to monitor applications deployed to TAP for their software supply chain in progress and their metadata using Backstage, an open source project

- **Supply Chain Choreographer** – a tool based on an open source project named Cartographer, allowing you to define a preapproved path to production for applications based on the kind of application

- **Cloud Native Runtime** – a serverless container platform for Kubernetes based on an open-source project named Knative

Figure 13.1 shows the Application Accelerator module as a part of the broader TAP user interface:

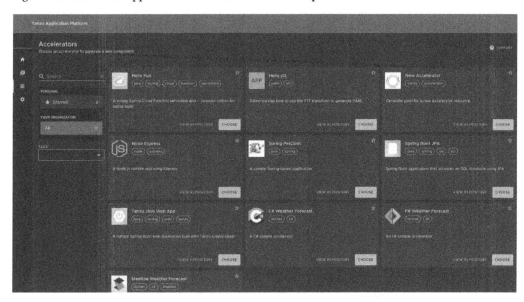

Figure 12.1 – Application templates from the TAP GUI

The next in the list is a bundle of products clubbed together for end-to-end Kubernetes operations – VMware Tanzu for Kubernetes Operations. Let's take a look at it.

VMware Tanzu for Kubernetes Operations

This product bundle is made for Kubernetes platform owners, giving them all the capabilities required to manage large Kubernetes foundations. It includes several products that we have covered in detail in this book already, as listed in the following points:

- **Tanzu Kubernetes Grid** – the multi-cloud Kubernetes distribution
- **Tanzu Mission Control** – a single pane of glass to control and manage hundreds of Kubernetes clusters
- **VMware Aria operations for Applications** – the full-stack observability tool providing visibility into infrastructure to applications deployed anywhere
- **Tanzu Service Mesh** – the cross-cluster networking tool for secure app-to-app traffic
- **NSX Advanced Load Balancer** – the scalable and software-based application load balancer for ingress traffic control with the security features of a web application firewall

Tanzu Data Solutions

Tanzu includes a set of data services that can be deployed on virtual machines, as Cloud Foundry managed services, and as containers running on Kubernetes. This bundle includes enterprise-supported offerings of the following open source products:

- **GemFire** – a reliable distributed cache with reliable data storage based on the open source Apache Geode project

- **RabbitMQ** – a data queuing solution for asynchronous communication based on its open source distribution

- **Tanzu SQL** – including MySQL and PostgreSQL RDBMS database offerings based on their open source distributions

- **Greenplum** – a data analytics platform based on an open source PostgreSQL database

We have not covered this product set in the book, as it simply includes the open source products with minor changes and enterprise support.

VMware Spring Runtime

We briefly covered this product offering in *Chapter 2, Developing Cloud-Native Applications*. It includes commercial enterprise support for several open source application development frameworks, runtimes, and tools as listed in the following points:

- Several Spring Framework projects, including but not limited to Spring Boot, Spring Security, and many others

- Spring Cloud services:

  - Spring API server

  - Configuration server

  - Spring Cloud Kubernetes

  - Spring Cloud Stream

  - Spring Cloud Data Flow

- Open JDK – a flavor of Java runtime that is open source

- The Apache Tomcat server

In addition to these Tanzu product bundles, there are other products that we have not covered in this book in detail. Let's review them in the next section of the chapter.

Tanzu beyond this book

The scope of this book was to cover those Tanzu products that play key roles in building, running, and managing modern containerized applications on top of Kubernetes platforms. However, the Tanzu ecosystem has a few more offerings that we have not covered so far, as listed in the following sub-sections.

Tanzu Labs

Tanzu Labs is a professional services wing under the Tanzu umbrella. It covers services that are outcome focused and consist of short-term engagements of typically 6 to 12 weeks based on the scope of work. These services include consultation from highly trained architects and engineers in the respective area of work. Each engagement typically involves a combination of two full-time engineers or architects and one product manager. These professionals work with the customer's team members following the practice of extreme programming. This way, the customer has the required, trained members to take ownership of the application or platform on which they are working without needing any more help after this short-term engagement. Tanzu Labs offers services in the following categories:

- **Building new apps** – an engagement to build a business-critical app from scratch to take the right first step

- **Modernizing an existing app** – an engagement to modernize an existing legacy system to make it cloud-native, typically re-architecting it with modern tools, technologies, and platforms

- **Building a platform** – an engagement to deploy a production-grade platform using multiple Tanzu products

- **Transforming data** – an engagement around data service optimization, architecture, and deployment for the data products included in the Tanzu Data Services bundle

- **Analyzing an application portfolio** – an engagement to analyze hundreds of applications in an organization to build a cloud migration plan for them to define one of the re-host, re-platform, relocate, re-architect, retain, and retire strategies for them

Tanzu Labs was also known as Pivotal Labs prior to the acquisition of Pivotal by VMware. This way, Tanzu Labs has a successful track record in providing these services for very large enterprises. Learn more about Tanzu Labs here: `https://tanzu.vmware.com/labs`.

Tanzu Application Service

Tanzu Application Service was formerly known as Pivotal Cloud Foundry, a commercially supported version of the open source Cloud Foundry platform. Like Kubernetes, Cloud Foundry is also a cloud-agnostic container platform that provides much-enhanced developer productivity. With just a small command, `cf push`, we can deploy a cloud-native application from a source to a container without any custom automation. Pivotal Cloud Foundry, and hence Tanzu Application Service, is a battle-tested technology that runs over a million containers, hosting mission-critical applications across the globe. The rise and popularity of Kubernetes overshadowed this technology but otherwise, it is a very mature and reliable container platform suitable for enterprises of any size and scale. Learn more about Tanzu Application Service here: `https://tanzu.vmware.com/application-service`.

Azure Spring Apps

In partnership, Microsoft and VMware launched a microservice platform on Azure that is powered by the Spring Framework, Tanzu, and AKS. It also supports the deployment of .NET Core and applications developed in other languages. Under the hood, this offering uses TAP capabilities (which we learned about in *Chapter 8, Enhancing Developer Productivity with Tanzu Application Platform*) on top of AKS. It provides Azure native customers with a familiar user interface to deploy their cloud-native applications using Tanzu's developer productivity and Azure's scalable infrastructure. Learn more about Azure Spring Apps here: `https://tanzu.vmware.com/azure-spring-apps`.

Concourse

Concourse is a commercially supported but very mature, open source, and lightweight **continuous integration/continuous deployment (CI/CD)** automation platform under the Tanzu umbrella. With its horizontally scalable design and ability to run its workers in parallel as containers, it is used in very large CI/CD automation platforms in some big enterprises. *Figure 13.2* shows one such complex pipeline on Concourse's web console. This platform is fully driven by YAML-based declarative configuration, adhering to the GitOps model and monitored using its sleek web-based UI to check the configured pipeline status. Learn more about Concourse here: `https://tanzu.vmware.com/concourse`.

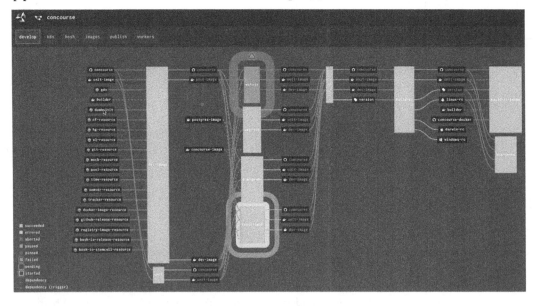

Figure 12.2 – A complex Concourse pipeline

Figure 13.2 shows a complex CI/CD pipeline deployed on Concourse. The texts of the labels and different pipeline stages are not readable, as the idea here is to show how a complex pipeline would look on the Concourse GUI. The black labels on the left-hand side of a box show inputs of data and configuration to the corresponding pipeline tasks. The similar labels on the right show the output of the corresponding tasks, which can be input for the next task(s). The columns of boxes show different stages executed in the sequence from left to right. The boxes stacked in the same column show the tasks that could be executed in parallel in the same stage. A red box shows an unsuccessful task. A box with a yellow border shows the task currently being executed. The green boxes show either completed or pending tasks in reference to the current task being executed.

With this, we covered some level of detail for all the offerings in the Tanzu portfolio. And now, let's make more sense of that by covering an end-to-end picture involving most of the offerings of Tanzu.

The end-to-end picture

We have covered over 20 different products in the Tanzu portfolio in various levels of detail in this book so far. Now is the time to club some of them together on a single page to understand how they work in tandem to provide a production-grade ecosystem for your applications with different levels of needs. *Figure 13.3* shows the key products in the Tanzu portfolio that work together to provide a comprehensive Kubernetes platform that helps implement required DevSecOps practices around modern applications:

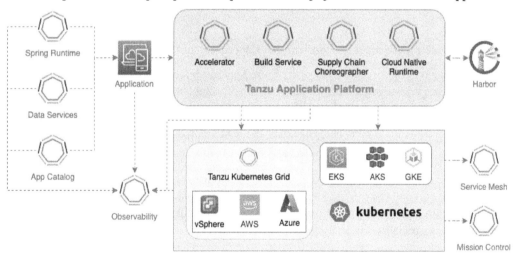

Figure 12.3 – The end-to-end picture of Tanzu

The following points describe *Figure 13.3*:

- An application could use the following:

  - VMware Spring Runtime for OpenJDK, Tomcat, and the Spring Framework as required

  - Tanzu Data Services for required backend data storage

  - VMware Application Catalog for any open source software requirement

- An application could be hosted on TAP, which uses several components internally to pave the application's path to production

- TAP could use Harbor as a container image repository to place application-built images from Tanzu Build Service and use those images to deploy the containers

- TAP could be deployed on one of the four Kubernetes distributions including the following:

  - Tanzu Kubernetes Grid – deployed on vSphere for on-premises deployment and AWS or Azure for a public cloud deployment

  - **EKS**

  - **AKS**

  - **GKE**

- Tanzu Service Mesh could connect required Kubernetes clusters for secure app-to-app traffic

- Tanzu Mission Control could manage the entire Kubernetes footprint for various types of policy control, life cycle management, backup/restore, and compliance audits

- VMware Aria operations for Applications could collect telemetry data from all the sources to provide end-to-end visibility of the health of all the monitored components at a central location

In addition to this deployment option, there are also other possible ways to deploy an application in containers using relevant Tanzu products. One of them is to use the custom CI/CD tools instead of TAP and deploy the containers in the Kubernetes clusters deployed by Tanzu Kubernetes Grid. The relevance of all other products would remain the same. Additionally, the custom CI/CD flow could be deployed using Concourse.

After understanding the big summarized picture, let's now understand what it means to get everything from a single vendor such as VMware in this case.

Pros and cons of a single-vendor solution

Using an end-to-end solution such as Tanzu or a single public cloud has many advantages but also some disadvantages. An enterprise should consider these points carefully and create a technology platform adoption plan accordingly. The following points outline the pros and cons of using a single-vendor solution containing multiple products.

Pros

- There is a big shift in the field of software to use open source technologies to avoid vendor lock-ins and gain the benefits of community collaboration. However, picking the right open source technology that is mature and supported is often a challenging and time-consuming process. That's where a comprehensive technology platform such as Tanzu could be a big help, as it uses many cherry-picked, open source tools under the hood. This will save considerable effort for an enterprise, along with giving it the confidence to use open source technologies heavily.

- When an enterprise picks different software solutions from different vendors or the open source community, it can be difficult to make those tools work together as intended. It is possible that nobody has ever tried and tested them in a single solution and that we will get unpleasant surprises as we move forward. Even if we make them work together at a certain point, there is no surety that upcoming versions of those tools from heterogeneous sources will keep working in harmony. That's where an end-to-end solution such as Tanzu could be helpful, as all the products in the portfolio are designed and ensured to work together. Otherwise, there will be no clarity on which version of one product is compatible with one version of another product.

- A large enterprise committing to buy multiple products from a single vendor could potentially get huge discounts on the license cost versus using different products from different vendors.

- In case of a big production outage, it would be helpful for an enterprise to have products from a single vendor in the solution rather than multiple vendors, or purely community-supported open source products. Having products from one vendor ensures required accountability without any questions of potential product compatibility issues.

- Using a platform such as Tanzu could result in operational efficiency gain, as there is a similar user experience while working on different tools. Common toolsets and user interfaces speed up the learning curve and the automation efforts around the platform. Eventually, this results in the agility and speed of modernization efforts in organizations using single-vendor solutions.

Cons

- An enterprise using multiple vendors for different products that are based on open source technologies could keep itself from heavily relying on one vendor for its success. A single-vendor solution could result in potential vendor lock-ins if the vendor does not use open source tools that are also supported by other vendors.

- Having all the products from a single vendor is like putting all eggs in one basket. If the basket is dropped, all the eggs will break. To avoid this situation, using more than one platform with the ability to move the applications around could be a useful mitigation strategy.

- Having an end-to-end solution could possibly mean having a set of products that lacks some capabilities available in similar products from different vendors. It is possible that an enterprise may need to weigh its choices based on the need-to-have capabilities rather than good-to-have ones. Every software vendor has a few shinier and more mature products than others, so having multiple products from the same vendor could mean missing out on some of the great alternatives available on the market for a subset of the products in the portfolio.

In software solution architecture, there is no silver bullet for any problem. Decisions are taken based on what is more important for an enterprise, considering the benefits of one approach, and how the risks are planned to be addressed for the drawback of the same approach. Using an end-to-end platform such as Tanzu is no different in this case.

The future of Kubernetes

While nobody can predict the future for sure, we can forecast a few things about Kubernetes based on the present technology adoption data and the pipeline work in progress in this ecosystem. As we learned in *Chapter 7, Orchestrating Containers across Clouds with Tanzu Kubernetes Grid,* in relation to Tanzu Kubernetes Grid, the adoption of Kubernetes has reached unprecedented heights in the last 2 years. It has become home to most newly developed and many legacy applications. It is the new middleware that abstracts the underlying infrastructure, providing true application portability that was not available before the emergence of the containerization of applications. Containerization changed the way we architect and think about software development. This containerization technology spanned various platforms in the past few years, including Docker Swarm, Mesosphere, AWS Elastic Container Service, Cloud Foundry, and Kubernetes. While all of them are mature and reliable technologies, Kubernetes is undoubtedly the most popular choice in this space. With the rise of running software as containers, most software vendors package their products as container images that can run on Kubernetes.

However, this is all a present-case scenario. Considering how things are moving today, we can attempt to predict Kubernetes' tomorrow as outlined in the following points:

- As the layer of virtual infrastructure has become the new default infrastructure bottom for most applications, Kubernetes may take its place and become the default infrastructure layer for applications going forward. In a way, Kubernetes could become ubiquitous.

- In the age of microservices and the **Internet of Things** (**IoT**), applications have become very small and suitable for running in containers. At the same time, deploying them closer to the point of usage is deemed more optimal. With the spur of applications needing to run at edge locations such as stores, cellphone towers, ships, airplanes, homes, and branch offices, we would see the use cases of Kubernetes deployments at the edge increasing in the future. This will require capabilities such as Tanzu Kubernetes Grid for on-premises deployment, Tanzu Mission Control to centrally manage hundreds of clusters, and VMware Aria operations for Applications to monitor those clusters, running apps on them and the underlying infrastructure centrally.

- With the increasing use of Kubernetes, aspects such as TAP would become mainstream tools for hiding Kubernetes clusters as below-value-line concerns. Soon, Kubernetes cluster management would be considered as an overhead. Developers would be spending lesser time writing Kubernetes-specific configurations to deploy their apps on Kubernetes clusters and spending more time enriching the application functionality. As developers do not think of any server's operating system-specific details before deploying their apps, they would similarly also stop thinking about the Kubernetes layer in near future.

- Before the emergence of public cloud platforms about a decade back, virtual servers were being treated as pets having names, receiving love and care, and having a long life. However, the API-based cloud infrastructure-as-code has made it so easy and required to create new servers in minutes that virtual servers are like cattle. The same is the case with Kubernetes clusters, which are considered pets today. With the need for more clusters, they may also be treated like cattle down the line.

- With its ever-growing popularity and the collaboration of its vibrant community, we may see a lot of new capabilities added to Kubernetes in the future, solving many open problems around it that we need to leverage external tools to solve today.

Kubernetes is here to stay for a long time but the way we care about it today could change tomorrow – and it should change because no business gets more revenue simply on the basis of running a healthy Kubernetes cluster. Kubernetes is just a layer of infrastructure enabling our revenue-generating apps to thrive.

What is next for Tanzu?

VMware created a new modern applications business unit in 2020 by clubbing the relevant products built internally and acquired from Pivotal, Bitnami, Heptio, and Wavefront. It is a well-thought-out portfolio covering most of the major areas of building, running, and managing modern apps in multi-cloud environments. All the products in this portfolio have good long roadmaps to add more capabilities and support for different cloud infrastructure providers. However, with the recent announcement of Broadcom Inc acquiring VMware, there are changes expected in how these products are grouped, along with other VMware and Broadcom's existing products. More details will unfold in the future post-acquisition. In any case, we believe that the value these products bring to the table would not change even if their packaging or names changed in the future, so we also hope that the knowledge gained in this book will help its readers nevertheless.

> **Important note**
> Anything mentioned in the following list is a forward-looking statement predicting the future product roadmap. It is possible that some or all the listed items could be dropped from the roadmap without notice.

The following list includes some of the existing work streams for different Tanzu products:

- Tanzu Mission Control and Tanzu Service Mesh could have an on-premises deployment variant in addition to the existing cloud-based **software-as-a-service** (**SaaS**) deployments to serve air-gapped deployments

- Tanzu Mission Control could have more capabilities to install and configure third-party software on the Kubernetes clusters under its management

- VMware Aria operations for Applications could also have a log aggregation capability, along with metrics, histograms, and span logs, to have everything in one place as far as full-stack observability is concerned

- Tanzu Kubernetes Grid and TAP could support more underlying cloud platforms, including OpenShift, Google Cloud Platform, and Oracle Cloud

- The Application Live View component of TAP could support custom modules developed by Tanzu customers as the plugins for the underlying open source Backstage project

Let's wrap up this chapter with a quick summary.

Summary

To conclude this book, we covered different commercial bundles available today in the market that club together different products in this book. We also learned about other products and bundles available under the Tanzu umbrella, including Tanzu Data for different types of backend data store requirements, Azure Spring Runtime to best utilize Azure cloud for Spring applications, and Tanzu Application Service, which is a commercial offering of Cloud Foundry and Concourse – a lightweight CI/CD automation tool. We also learned about the end-to-end deployment picture of Tanzu deployment using key products we covered in this book. We also learned what it means to embrace multiple products from a single vendor. While there are some drawbacks to picking a one-stop shop for many of your needs, there are also considerable benefits to doing so. Finally, we predicted the future of Kubernetes and Tanzu based on their current conditions.

With this, we conclude this book. Thank you for your learning journey with us so far! We hope that you learned about the Tanzu portfolio as per your expectations. As was said previously in the book, we have just scratched the surface. For more learning resources, refer to the *Appendix* section of the book.

Appendix

In this section, we have included additional learning resources around Tanzu and the references to create a Kubernetes cluster for the hands-on work required in this book.

Additional learning resources from VMware

The following are some good learning references to further explore Tanzu:

- The official VMware Tanzu product documentation: `https://docs.vmware.com/allproducts.html#section`
- VMware provided on-demand training courses for developers and operators: `https://tanzu.vmware.com/education`
- Hands-on workshops for Tanzu: `https://tanzu.vmware.com/developer/workshops/`
- Hands-on labs and courses related to Tanzu and building modern apps: `https://modernapps.ninja/`
- Kubernetes courses: `https://kube.academy/`
- Tanzu channel on YouTube: `https://www.youtube.com/c/VMwareTanzu`

Different ways to create a Kubernetes cluster

Most of the chapters in the book list having a Kubernetes cluster as one of the prerequisites to follow the instructions in each chapter. While we have learned about three different approaches to creating Tanzu clusters in this book as referenced in the following points, there are also various other ways to get your Kubernetes cluster up and running with minimal effort. This section lists some of the many ways to create a Kubernetes cluster.

Creating Tanzu Kubernetes Grid clusters

The following list include ways to create Tanzu Kubernetes Grid clusters:

- **Tanzu Kubernetes Grid**: In *Chapter 7, Orchestrating Containers across Clouds with Tanzu Kubernetes Grid*, we described how you can use Tanzu Kubernetes Grid to create a managed enterprise-ready Kubernetes cluster.

- **Tanzu Mission Control**: Tanzu Mission Control makes it very simple and straightforward to create a Kubernetes cluster. We covered it in detail in *Chapter 9, Managing and Controlling Kubernetes Clusters with Tanzu Mission Control*.

- **vSphere with Tanzu**: If you're an existing vSphere cluster, you may have the ability to create clusters with a **Supervisor Cluster** already built into your virtual infrastructure. You can learn more here: `https://docs.vmware.com/en/VMware-vSphere/7.0/vmware-vsphere-with-tanzu/GUID-3B2102E6-D9AA-4FE6-B3AA-60B450BE8491.html`.

Creating non-Tanzu Kubernetes clusters

The following list include ways to create non-Tanzu Kubernetes clusters on public clouds, local desktop environment, and on the OpenShift platform:

- Creating a local Kubernetes cluster using Docker Desktop: `https://docs.docker.com/desktop/kubernetes/`

- Creating a local Minikube cluster: `https://minikube.sigs.k8s.io/docs/start/`

- Creating an AWS **Elastic Kubernetes Service** (**EKS**) cluster: `https://docs.aws.amazon.com/eks/latest/userguide/create-cluster.html`

- Creating a **Google Kubernetes Engine** (**GKE**) cluster: `https://cloud.google.com/kubernetes-engine/docs/how-to/creating-a-zonal-cluster`

- Creating an **Azure Kubernetes Service** (**AKS**) cluster: `https://docs.microsoft.com/en-us/azure/aks/learn/quick-kubernetes-deploy-portal?tabs=azure-cli`

- Creating a Red Hat OpenShift cluster on IBM Cloud: `https://cloud.ibm.com/docs/openshift?topic=openshift-clusters&interface=ui`

Index

Other Books You May Enjoy

If you enjoyed this book, you may be interested in these other books by Packt:

Learning DevOps - Second Edition

Mikael Krief

ISBN: 9781801818964

- Understand the basics of infrastructure as code patterns and practices
- Get an overview of Git command and Git flow
- Install and write Packer, Terraform, and Ansible code for provisioning and configuring cloud infrastructure based on Azure examples
- Use Vagrant to create a local development environment
- Containerize applications with Docker and Kubernetes
- Apply DevSecOps for testing compliance and securing DevOps infrastructure
- Build DevOps CI/CD pipelines with Jenkins, Azure Pipelines, and GitLab CI
- Explore blue-green deployment and DevOps practices for open sources projects

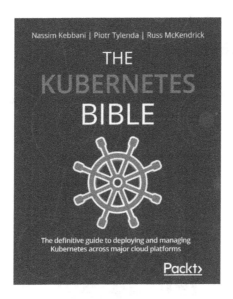

The Kubernetes Bible

Nassim Kebbani, Piotr Tylenda, Russ McKendrick

ISBN: 9781838827694

- Manage containerized applications with Kubernetes
- Understand Kubernetes architecture and the responsibilities of each component
- Set up Kubernetes on Amazon Elastic Kubernetes Service, Google Kubernetes Engine, and Microsoft Azure Kubernetes Service
- Deploy cloud applications such as Prometheus and Elasticsearch using Helm charts
- Discover advanced techniques for Pod scheduling and auto-scaling the cluster
- Understand possible approaches to traffic routing in Kubernetes

Packt is searching for authors like you

If you're interested in becoming an author for Packt, please visit `authors.packtpub.com` and apply today. We have worked with thousands of developers and tech professionals, just like you, to help them share their insight with the global tech community. You can make a general application, apply for a specific hot topic that we are recruiting an author for, or submit your own idea.

Share Your Thoughts

Now you've finished *DevSecOps in Practice with VMware Tanzu*, we'd love to hear your thoughts! Scan the QR code below to go straight to the Amazon review page for this book and share your feedback or leave a review on the site that you purchased it from.

`https://packt.link/r/1-803-24134-9`

Your review is important to us and the tech community and will help us make sure we're delivering excellent quality content.

Download a free PDF copy of this book

Thanks for purchasing this book!

Do you like to read on the go but are unable to carry your print books everywhere?

Is your eBook purchase not compatible with the device of your choice?

Don't worry, now with every Packt book you get a DRM-free PDF version of that book at no cost.

Read anywhere, any place, on any device. Search, copy, and paste code from your favorite technical books directly into your application.

The perks don't stop there, you can get exclusive access to discounts, newsletters, and great free content in your inbox daily

Follow these simple steps to get the benefits:

1. Scan the QR code or visit the link below

https://packt.link/free-ebook/9781803241340

2. Submit your proof of purchase
3. That's it! We'll send your free PDF and other benefits to your email directly